Other titles published in this series (continued):

Soft Sensors for Monitoring and Control of Industrial Processes
Luigi Fortuna, Salvatore Graziani, Alessandro Rizzo and Maria G. Xibilia

Adaptive Voltage Control in Power Systems
Giuseppe Fusco and Mario Russo

Advanced Control of Industrial Processes
Piotr Tatjewski

Process Control Performance Assessment
Andrzej W. Ordys, Damien Uduehi and Michael A. Johnson (Eds.)

Modelling and Analysis of Hybrid Supervisory Systems
Emilia Villani, Paulo E. Miyagi and Robert Valette

Process Control
Jie Bao and Peter L. Lee

Distributed Embedded Control Systems
Matjaž Colnarič, Domen Verber and Wolfgang A. Halang

Precision Motion Control (2nd Ed.)
Tan Kok Kiong, Lee Tong Heng and Huang Sunan

Optimal Control of Wind Energy Systems
Julian Munteanu, Antoneta luliana Bratcu, Nicolaos-Antonio Cutululis and Emil Ccangă

Identification of Continuous-time Models from Sampled Data
Hugues Gamier and Liuping Wang (Eds.)

Model-based Process Supervision
Arun K. Samantaray and Belkacem Bouamama

Diagnosis of Process Nonlinearities and Valve Stiction
M.A.A. Shoukat Choudhury, Sirish L. Shah and Nina F. Thornhill

Magnetic Control of Tokamak Plasmas
Marco Ariola and Alfredo Pironti

Real-time Iterative Learning Control
Jian-Xin Xu, Sanjib K. Panda and Tong H. Lee

Deadlock Resolution in Automated Manufacturing Systems
ZhiWu Li and MengChu Zhou

Model Predictive Control Design and Implementation Using MATLAB®
Liuping Wang

Predictive Functional Control
Jacques Richalet and Donal O'Donovan

Fault-tolerant Flight Control and Guidance Systems
Guillaume Ducard

Fault-tolerant Control Systems
Hassan Noura, Didier Theilliol, Jean-Christophe Ponsart and Abbas Chamseddine

Detection and Diagnosis of Stiction in Control Loops
Mohieddine Jelali and Biao Huans (Eds.)

Stochastic Distribution Control System Design
Lei Guo and Hong Wang

Dry Clutch Control for Automotive Applications
Pietro J. Dolcini, Carlos Canudas-de-Wit and Hubert Béchart

Advanced Control and Supervision of Mineral Processing Plants
Daniel Sbárbaro and René del Villar (Eds.)

Active Braking Control Design for Road Vehicles
Sergio M. Savaresi and Mara Tanelli

Active Control of Flexible Structures
Alberto Cavallo, Giuseppe de Maria, Ciro Natale and Salvatore Pirozzi

Induction Motor Control Design
Riccardo Marino, Patrizio Tomei and Cristiano M. Verrelli

Fractional-order Systems and Controls
Concepcion A. Monje, YangQuan Chen, Blas M. Vinagre, Dingyu Xue and Vincente Feliu

Model Predictive Control of Wastewater Systems
Carlos Ocampo-Martinez

Wastewater Systems
Carlos Ocampo-Martinez

Tandem Cold Metal Rolling Mill Control
John Pitter and Marwan A. Simaan

Advances in Industrial Control

Advances in Industrial Control is a series of monographs and contributed titles focusing on the applications of advanced and novel control methods within applied settings. This series has worldwide distribution to engineers, researchers and libraries.

The series promotes the exchange of information between academia and industry, to which end the books all demonstrate some theoretical aspect of an advanced or new control method and show how it can be applied either in a pilot plant or in some real industrial situation. The books are distinguished by the combination of the type of theory used and the type of application exemplified. Note that "industrial" here has a very broad interpretation; it applies not merely to the processes employed in industrial plants but to systems such as avionics and automotive brakes and drivetrain. This series complements the theoretical and more mathematical approach of Communications and Control Engineering.

Indexed by SCOPUS and Engineering Index.

Proposals for this series, composed of a proposal form (please ask the in-house editor below), a draft Contents, at least two sample chapters and an author cv (with a synopsis of the whole project, if possible) can be submitted to either of the:

Series Editor

Professor **Michael J. Grimble**:
Department of Electronic and Electrical Engineering, Royal College Building, 204 George Street, Glasgow G1 1XW, United Kingdom
e-mail: m.j.grimble@strath.ac.uk

or the

In-house Editor

Mr. **Oliver Jackson**:
Springer London, 4 Crinan Street, London, N1 9XW, United Kingdom
e-mail: oliver.jackson@springer.com

Proposals are peer-reviewed.

Publishing Ethics

Researchers should conduct their research from research proposal to publication in line with best practices and codes of conduct of relevant professional bodies and/or national and international regulatory bodies. For more details on individual ethics matters please see: https://www.springer.com/gp/authors-editors/journal-author/journal-author-helpdesk/publishing-ethics/14214.

Balázs Németh · Péter Gáspár

Control of Variable-Geometry Vehicle Suspensions

Design and Analysis

Springer

Balázs Németh
Systems and Control Laboratory
Institute for Computer Science and Control (SZTAKI)
Eötvös Loránd Research Network (ELKH)
Budapest, Hungary

Péter Gáspár
Systems and Control Laboratory
Institute for Computer Science and Control (SZTAKI)
Eötvös Loránd Research Network (ELKH)
Budapest, Hungary

ISSN 1430-9491 ISSN 2193-1577 (electronic)
Advances in Industrial Control
ISBN 978-3-031-30536-8 ISBN 978-3-031-30537-5 (eBook)
https://doi.org/10.1007/978-3-031-30537-5

Mathematics Subject Classification: 93-10, 93B36, 93B51, 93C85, 93D05

This Springer imprint is published by the registered company Springer Nature Switzerland AG
The registered company address is: Gewerbestrasse 11, 6330 Cham, Switzerland

Series Editor's Foreword

Control engineering is viewed differently by researchers and engineers that must design, calibrate, implement, and maintain control systems. Researchers often develop algorithms for general control problems with a well-defined mathematical basis; engineers have more immediate concerns over the limitations of equipment, quality of control, safety, security, and system downtime. The monograph series on *Advances in Industrial Control* attempts to bridge this divide by encouraging the adoption of advanced control techniques when they offer real benefits.

The rapid development of new control theory, techniques, and technology has an impact on all areas of engineering applications. This series of monographs and contributed volumes has a focus on applications that often stimulate the development of new more general control paradigms. This is desirable if the different aspects of the "control design" problem are to be explored with the same dedication that "control analysis and synthesis" problems have received. The series enables researchers to introduce new ideas motivated by challenging problems in the applications of interest. It raises awareness of the various benefits that advanced control can provide while explaining the challenges that can arise.

This *Advances in Industrial Control* series monograph is concerned with variable-geometry vehicle suspensions and the opportunities the technology provides. The control of these systems involves both model-based advanced controls and data-driven learning systems. The use of artificial intelligence (AI) and machine learning in industrial applications has generated great interest and the variable-geometry-suspension problem is a good candidate for introducing learning features.

The basic ideas and history of variable-geometry suspension systems are reviewed in Chap. 1. Possible benefits, such as increased maneuvrability, reduction of tire wear, and minimization of vehicle roll angle, are described. Chapter 2 covers the popular linear parameter-varying (LPV)-based modeling approach applied to variable-geometry suspension systems. The physical system equations are introduced, and parameter variations are described.

Chapter 3 turns to the control problem including the LPV modeling of electro-hydraulic actuators. The parameters are likely to contain some level of uncertainty, so the robust control design of actuator positioning controls is discussed. The simulation of a suspension control system illustrates the ideas.

Chapter 4 considers the nonlinear dynamics like the characteristics of lateral tire forces. Model formulation for nonlinear lateral vehicle dynamics is included and an LPV approach to trajectory tracking is explained. A quadratic LPV performance problem that determines the parameter-varying controller is established. The solution can be found using a linear matrix inequality approach. A design example for a vehicle that has independent suspension on the front and the rear axles is included.

To achieve independent steering action, Chap. 5 considers the development of control-oriented models for variable-geometry suspension systems. In this case, the motion dynamics of the suspension system and the orientation of the wheel are considered. Chapter 6 concerns hierarchical control design and the coordination of steering control and torque-vectoring for electric vehicles. The design of the control law for an electro-hydraulic suspension actuator is also described. The hierarchical structure involves different levels for lateral control, wheel steering control, and suspension control.

Chapter 7 considers a coordinated control strategy for variable-geometry suspension systems using a polynomial analysis approach. It involves the coordinated control of independent steering and torque-vectoring functions, which requires quite a complex system architecture. Chapter 8 is more practical and covers the tracking-control-law problem with implementation on a suspension test bed that includes nonlinearities and uncertainties.

Chapter 9 concerns a data-driven framework for variable-geometry suspension control. An LPV model is used and a parameter-varying controller is determined. A hardware-in-the-loop simulation is used, illustrating the path following capability. It is always a problem to obtain relevant data to test such algorithms, but in this case the authors have generated their own dataset from measurements on a suspension test bed.

Chapter 10 turns to guaranteeing the performance of a suspension system using a neural network. The use of machine learning in such applications is much in vogue currently, and it is good to have concrete examples of how the approach may be used. The control design of variable-geometry suspension systems with learning methods is considered in Chap. 11. The aim is to limit the lateral error in trajectory tracking, where independent steering is used. The second example included involves a driver-in-the-loop with performance guarantees.

The rapid change in the automotive industry brought about by the move to electric and hydrogen-fuelled vehicles has provided opportunities for the inclusion of more advanced controls for all aspects of road vehicle operation. New actuator and sensor systems provide the mechanisms for improved performance, and this monograph illustrates the flexibility and performance benefits achievable in the particular area of variable-geometry suspensions. The authors have provided a valuable text for anyone working on vehicle dynamics and control. It touches on the "hot" topic of data-driven AI-based algorithms for such applications. The monograph should be useful to vehicle-control-design engineers, researchers in dynamics and control, and engineering students hoping to join this automotive-led industrial revolution.

Glasgow, UK
December 2022

Michael J. Grimble

Preface

The variable-geometry suspension is a novel type of smart actuator with which several functionalities in automated vehicles can be achieved, e.g., independent steering for trajectory tracking or improving traveling comfort. For all functionalities, the theoretical backgrounds of modeling, control design and analysis, together with their practical implementation solutions, are detailed. This book focuses on two important types of variable-geometry vehicle suspensions, and, thus, the theoretical results through simulation examples and Software-in-the-Loop, Hardware-in-the-Loop and Driver-in-the-Loop scenarios are confirmed.

Concerning the variable-geometry suspension system, the book is divided into three main parts. First, modeling and robust Linear Parameter-Varying (LPV) synthesis methods for achieving wheel tilting functionality in automated vehicles are provided. The advantage of wheel tilting control is that it meets the predefined performance specifications, such as the enhanced maneuvering capability of the vehicle, the reduction of tire wear and the minimization of the chassis roll angle. The proposed integrated design of the control system and suspension constructional parameters provides a new perspective in the research field.

Second, the creation of independent steering functionality with variable-geometry suspension is presented. Modeling methods for the control-oriented formulation dynamics of electro-hydraulic actuators, together with variable-geometry vehicle suspension dynamics and lateral vehicle dynamics, are proposed. The coordinated control of independent steering and torque vectoring in a hierarchical structure is yielded. In order to create an optimal coordination strategy, a nonlinear system analysis, i.e., Sum-of-Squares (SOS) method, is introduced. The operation of the control design on a unique test bed with Hardware-in-the-Loop scenarios is demonstrated.

Third, the emerging trend of automated vehicles, i.e., the integration of control design and learning methods to achieve enhanced suspension functions, is the focus of the book. Data-driven modeling and control design methods for variable-geometry suspension are provided, with the extension of the classical model-based robust control and LPV synthesis methods. The application of the proposed control for variable-geometry suspensions through Driver-in-the-Loop simulations is shown.

We recommend this book to specialists and M.Sc. and Ph.D. students in the fields of control theory and automotive systems. We also hope that this book will become a useful help to understand the most important control problems involved in variable-geometry suspensions, and also to find practical solutions to them.

Budapest, Hungary
January 2023

Balázs Németh
Péter Gáspár

Acknowledgements

This book is the result of many years of research in Institute for Computer Science and Control, Eötvös Loránd Research Network. The authors are grateful for the continuous support of the research activities, i.e., the work has been supported by the European Union within the framework of the National Laboratory for Autonomous Systems (RRF-2.3.1-21-2022-00002) and by the National Research, Development and Innovation Office (NKFIH) under OTKA Grant Agreement No. K 135512.

The authors thank Series Editor Prof. Michael J. Grimble for supporting the publication of the book. Special thanks also go to Springer editor Oliver Jackson and his team for their help during the preparation of the manuscript.

Contents

Chapter 1
Introduction

1.1 Motivation of Variable-Geometry Suspension Systems

In the last decade, several new focuses for research and development have arisen in the automotive industry. These focuses are on urban mobility and transport, alternative fuels and the electrification of the vehicle safety applications in co-operative systems; see EUCAR (1992). Self-driven vehicles, smart urban solutions and electric driveline systems have increased influence on the actual direction of automotive and transportation-oriented research. The motivation of these fields is the growing transportation demand in cities, where limitations on road traffic flow capacities can be found. Furthermore, the importance of electric vehicles with enhanced economic driveline control solutions is to reduce emissions in urban regions. The design of autonomous electric-driven vehicles with their connection to the highly automated transportation systems can help to overcome these issues. Nevertheless, it requires the use of V2X communication topologies, especially vehicle-to-vehicle and vehicle-to-infrastructure solutions. The requirements of small and light vehicle chassis for urban autonomous vehicles lead to the limitation of actuator numbers, but, similarly, the maneuverability of the vehicle cannot be reduced (Piyabongkarn et al. 2004). It motivates the coordination and design of automotive smart actuators, especially for urban vehicles, with which the requested vehicle control functionalities can be realized.

To cope with these requirements, variable-geometry vehicle suspension takes the form of an enhanced smart actuator that can achieve various vehicle control functionalities through the controlled tilting of the wheels, e.g., independent steering for trajectory tracking and the improvement of traveling comfort. Variable-geometry suspensions have various types with different constructions and control strategies. Some of these actuators can be integrated with other vehicle systems, e.g., in-wheel driving systems or electric steering systems. The integration of suspension with steering or driving can have three important motivations, such as the increased maneuvering capability, the guarantee of fault-tolerant vehicle operation and the advantageous impact on the various dynamical motions simultaneously.

B. Németh and P. Gáspár, *Control of Variable-Geometry Vehicle Suspensions*, Advances in Industrial Control, https://doi.org/10.1007/978-3-031-30537-5_1

Through the integration of variable-geometry suspension and steering systems, the maneuverability of the vehicle can be increased. For example, wheel tilting on the steered wheels can increase the lateral tire force of the vehicle, which can result in an increased region of maneuverability for the vehicle (Németh and Gáspár 2017). The integration of the active steering system and a variable-geometry suspension system is applied as a driver assistance system in vehicles. The purpose of trajectory tracking is to assist the driver and guarantee the road stability of the vehicle. While the driver performs a maneuver by using the steering wheel, an autonomous control system modifies the front wheel steering angle and the camber angle of the front wheels. In the operations of different control systems, interference or conflicts may occur between the control components. An integrated design of actuators is able to create a balance between them and handle possible conflicts.

The increased maneuvering capability for electric vehicles can be achieved through the integration of the variable-geometry vehicle suspension and some elements of the drivetrain system, e.g., in-wheel-structured electric motors (Németh et al. 2019). One of the main constructional benefits of in-wheel vehicles is the space-efficient passenger cabin design, which is essential for small city cars. The in-wheel electric drive poses new challenges in generating the differential yaw moment of the vehicle. The independent rear wheel steering to influence toe angle of the vehicle is proposed by Lee et al. (1999), Malvezzi et al. (2022) and Ronci et al. (2011). Torque vectoring requires fast torque generation with low error on each wheel by the in-wheel motors for achieving high-performance maneuvering capability; see, e.g., the methods of Castro et al. (2012), Shuai et al. (2013) and Wu et al. (2013). Being aware of the hub motor and the hydraulic brake system characteristics, the functionalities of regenerative braking and energy optimal torque distribution in the vehicle control can be implemented (Lin and Xu 2015; Wang et al. 2011b). Various variable-geometry-suspension-based independent steering systems require the independent driving of the wheels. Thus, the coordination of independent steering and driving of each wheel can result in increased maneuvering for the automated vehicle.

Automated and autonomous vehicles require control solutions, with which the safe motion of the vehicle with guaranteed performances under all scenarios is achieved. It requires fault-tolerant solutions in the vehicle control systems, e.g., the reconfiguration of the actuator interventions depending on the degradation of the subsystems. The reconfiguration strategy is strongly connected to the integration of the vehicle systems, with which the maneuvering capability of the vehicle can be enhanced.

A challenging area of fault-tolerant control in electric vehicles is the reconfiguration of the driving torques in the driven wheels. In the works of Hu et al. (2011) and Wang and Wang (2012), the design of a fault-tolerant control system has been proposed. It is based on the principle of the automatic reallocation of the control intervention from the faulted motors to the healthy wheels. Another fault-tolerant control method for electric vehicles with independently actuated drive on the wheels has been presented by Hu et al. (2015), in which the complete failure of the active steering system has been handled. The integration of variable-geometry suspension and an in-wheel driving system provides an alternative and effective solution to the problem of fault-tolerant control (Fényes et al. 2018). The contribution of the method is a reconfiguration strategy, with which the road holding and tracking performances of the automated electric vehicle can be guaranteed.

The improvement in the maneuvering capability and the guarantee of the fault-tolerant operation are important motivations for the integrated design of the vehicle control systems. Nevertheless, one of the most fruitful contributions of the integration is that the impact of the control systems on the various vehicle motions can be simultaneously enhanced. Through the integration with steering, in-wheel driving or braking systems, the performance level of vehicle control in the vehicle motion can be improved because variable-geometry suspension can influence the vertical motion of the chassis, the roll angle and the lateral motion of the vehicle.

For example, in the case of vehicle roll dynamics the roll center height has high importance. A solution for the reduction of chassis roll motion is the roll center height modification, i.e., its minimization. Since the position of the roll center is influenced by the front wheel camber angle, it can be effectively varied through variable-geometry vehicle suspension. Moreover, the motion of the wheel–road contact point in a lateral direction has relevance on the value of tire wear, which must also be considered in the control of the suspension system (Gough and Shearer 1956). Nevertheless, through variable-geometry suspension the impact of the unwanted up and down motion of the chassis on the contact point motion can also be compensated. Thus, under normal driving condition maneuvers, the steering system can have the role of path following, and through variable-geometry vehicle suspension can be focused on the fulfillment of another performance requirement, e.g., reduction of chassis roll motion and half-track change. Furthermore, through the variation of wheel tilting on the front wheels the vehicle yaw motion can also be influenced, and, thus, variable-geometry suspension can be applied for the reduction of path following error, e.g., it can be used for driver steering assistance purposes. It means that through a reconfiguration-based control system, various functionalities with the suspension can be achieved. For example, in the case of a high-speed cornering situation, the suspension control system can enforce path following. Nevertheless, it leads to the reduction of comfort-oriented and economy-based performance levels, such as increased chassis roll angle or tire wear. By achieving a trade-off in the control design, the conflict situation of different performance requirements can be handled; e.g., in the work of Németh and Gáspár (2013) a strategy with parameter-dependent weights on this problem has been proposed.

With the motivation of small-scaled urban vehicle and the integration of variable-geometry suspension with steering, in-wheel driving systems provide the perspective of an individual vehicle subsystem, similar to Active Wheel (Rajaie et al. 2019) and eCorner concepts (Heissing and Ersoy 2011; Doericht 2013). Its important functionality is the independent steering and driving for all wheels. Due to its individual nature, the concept owns the vision of a plug-and-play subsystem, as detailed below.

The purpose of independently controlled wheel steering is the improving of the vehicle lateral dynamics. A special concept of independent steering for variable-geometry suspension systems with the modification of toe angles on the rear wheels can be found in Lee et al. (1999). Examination of the applicability and effectiveness of independent steering in relation to heavy vehicles is presented by Ronci et al. (2011).

Validation and feasibility analysis of steering with differential drive torque assistance is found in the work of Wang et al. (2011a). In the concept of the individual subsystem, the variable-geometry suspension control is applied for independent steering control purposes. The advantage of this solution is that the controls of the variable-geometry suspensions are independent of each other. Thus, the steering of the left and right wheels can also be achieved independently.

The concept of the individual subsystem with independent steering and driving supersedes the perspective of the plug-and-play control solution. The advantage of the integrated control is that the control of the different subsystems can be individually designed and, simultaneously, a supervisory strategy provides the coordination of the subsystems. For example, in the case of adding a new component to the vehicle control, or replacing an old one with a new one, the architecture of the control or the physical system control can vary. Together with this modification, the controllers with conventional structures must be re-designed, which can have cost and time requirements. A solution addressed to this problem is the supervisory control structure, in which the modification of the components can lead only to the modification of the high-level supervisory algorithm. Thus, a strong motivation behind the research on control systems is to design plug-and-play control systems. It can provide the possibility of applying actuators and sensors, which the different vendors interchangeably on a core system are provided, while the minimum level of global performance requirements is guaranteed. The communication between the central control unit and control units on the local level is carried out through communication bus and interfaces (Pernebo and Hansson 2002). These interfaces form the basis of the design of the individual plug-and-play subsystems. However, in order to make the design plug-and-play-compatible, the baseline method must be modified. Its most important aspect is that fixed component architecture must be designed to handle varying characteristics of the devices, which can be replaced. This diversity imposes robustness requirements and a quest for non-conservative design for robust performance. A serious research effort was already done on the level of the individual vehicles, leading to the concept of plug-and-play control and the integrated multi-level solutions of the different subsystems; see, e.g., Stoustrup (2009). Plug-and-play control also has importance in the decentralized supervisory control; monitoring signals are used to examine the impact of control units, i.e., their operation in the modified context (Gáspár et al. 2017). In this concept, the design of the control systems is carried out at given operation points inside of a bounded range, and then, the individually designed controllers are independently implemented. The coordination among the controllers through their operation is guaranteed by the selection of the monitoring signal, with which the entire system is able to adapt to the actual environmental circumstances.

Finally, an important motivation of the research in the field of variable-geometry suspension is its simple structure, low energy consumption (Lee et al. 2005) and low cost (Evers et al. 2008a). Since it can be taken part of individual subsystems, the variable-geometry suspension can easily allocate in various wheels of the vehicle, with which additional functionalities for the vehicle can be achieved, e.g., independent steering in each wheel. The earlier suspension-controlled concepts can have limitations from economical or functional viewpoints. For example, active suspension systems can have an advantageous impact on the road holding capability and the rollover prevention of the chassis (Lugner 2019), but their constructions are expensive and the system has significant energy consumption (Martins et al. 2006). Semi-active suspension provides an achievable solution for the improvement of riding comfort and road holding, but the control design can be difficult due to the strict limitations in the intervention (Savaresi et al. 2010). Since in the case of small-scaled urban vehicles it is not recommended to apply several number of sensors and actuators due to economic reasons, it is requested to use vehicle systems, with which an increased number of functionalities can be achieved. The integration of variable-geometry suspension and in-wheel driving systems can be an appropriate choice to reach this goal.

1.2 Overview of Variable-Geometry Suspension Systems: Constructions and Control Methods

In the last decades, lots of papers on the topic of variable-geometry vehicle suspensions have been published, focusing on its modeling and control design. Since Sharp (1998) was provided one of the first reviews on the variable-geometry suspensions, several new solutions were developed. From the modeling side, formulation of the nonlinear dynamics of the McPherson type of suspension was described by Fallah et al. (2009) and Németh and Gáspár (2012). Linearization techniques were used by Hong et al. (1999), which can serve as a basis for the control design of active suspension. Paper Lee et al. (2009) proposed an optimization technique for the construction of the McPherson type suspension, i.e., the effects of king-pin angles, camber and caster on the motion of the performances of the suspension were considered. The impact of chassis roll motion on the steering of the vehicle was examined in the work of Habibi et al. (2008). Another analysis of the influence of vertical force distribution on the selection methodology of suspension components was provided by Braghin et al. (2008). Design methods for similar purposes in the context of double-wishbone suspension were provided by Sancibrian et al. (2010). One of the challenges of the design methods is to find an appropriate trade-off between different performance requirements; see Vukobratovic and Potkonjak (1999).

An important construction in the field of variable-geometry suspensions is the solution of direct wheel tilting; see Fig. 1.1. The concept of "carving" has been implemented in the vehicle Mercedes F400; see Schiehlen and Schirle (2006). In the presented construction, the wheel hub is divided into two parts. The first part is fixed to the arms of the double-wishbone suspension, and it contains a hydraulic cylinder. The second part of the hub is connected to the hydraulic cylinder and it rotates around a joint, which links together the two parts of the hub. Thus, in this construction the tilting of the wheel is independent of the positions of the arms, which means that high wheel tilting can be achieved. Although the presented solution is effective for wheel tilting, further functionality cannot be achieved. Direct wheel tilting can be

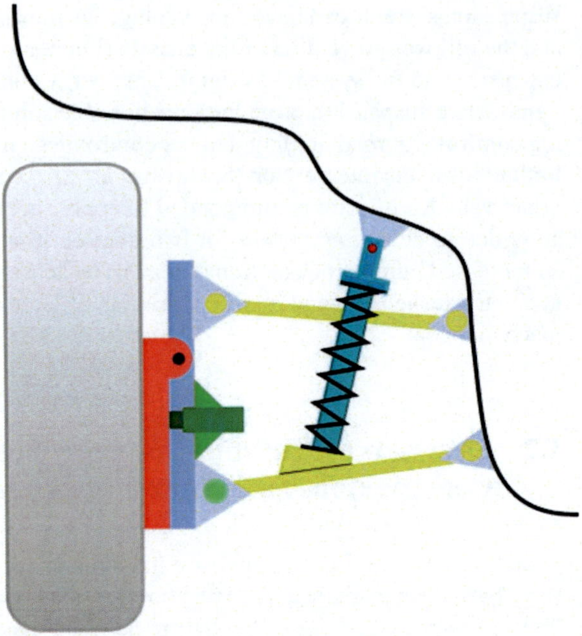

Fig. 1.1 Construction of direct wheel tilting

useful for vehicles, which have a low height in the roll center and are designed for high-speed motion, e.g., sports or racing cars.

Another important construction of the variable-geometry suspension system is illustrated in Fig. 1.2. The presented multilink type of suspension construction is implemented in Hyundai Sonata for the modification of the toe angle on the rear wheels (Lee 2010; Lee et al. 2006). A similar construction can also be found in Goodarzia et al. (2010). The goal of the toe angle control is the enhancement of driving stability. In the presented solution, a hydraulic actuator is fixed on a lower arm of the suspension, which provides the modification of the wheel position through the motion of an assist link. The assist link and the hydraulic cylinder are connected together through a lever with a joint, which is able to rotate. The presented construction is used for control purposes only with two modes, such as without the stroke of the cylinder or with maximum stroke. Thus, the toe angle can have two positions. Therefore, only one cylinder at the same time is actuated, with which the left or the right turning maneuvering can be improved.

The scheme of the construction for achieving suspension with variable roll center is found in Fig. 1.3. The goal of this solution is to modify the position of the lower arm of the suspension, with which the height of the chassis roll center can be influenced; see, e.g., the variation for the McPherson type of construction. The variation of the roll center can also be achieved as an additional effect of toe angle modification in the previous solution. Nevertheless, it can be used as an individual functionality for

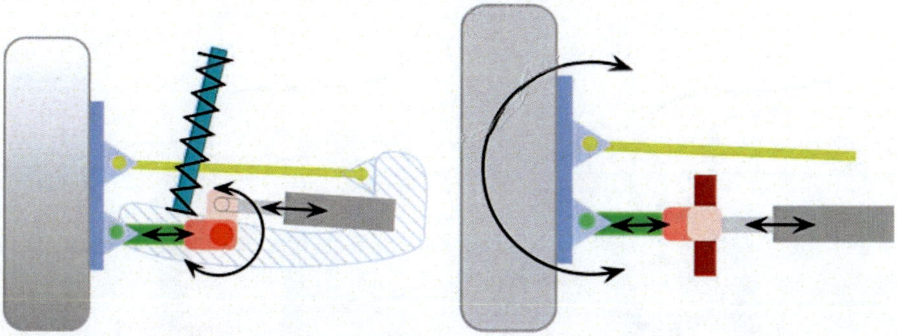

(a) Back view (b) Top view

Fig. 1.2 Construction of suspension for toe angle control

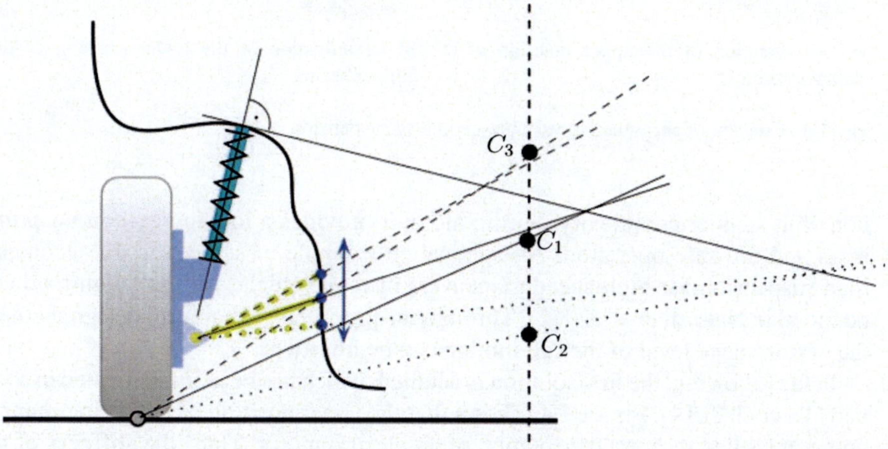

Fig. 1.3 Construction for achieving variable roll center

front suspensions (Lee and Han 2001; Lee et al. 2008). The goal of the intervention is to enhance the roll dynamics of the vehicle through the modification of the kinematic roll center.

The modification of the suspension geometry can be achieved through the modification of the position of the spring-damper unit. In the concept of series active variable-geometry suspension, there are two possible solutions. First, the connection point of the spring-damper unit and the chassis can be moved (Yu et al. 2021a; Arana et al. 2015; Anubi et al. 2013); see Fig. 1.4a. Second, the connection point of the damper/spring and the suspension arm can be moved in lateral direction (Watanabe and Sharp 1999); see Fig. 1.4b. Through the modification of the connection, the stiffness of the suspension can be modified, with which traveling comfort, road holding, pitch and roll dynamics can be improved. The purposes of these solutions are close to the active and semi-active suspension systems, but it requires cheaper construc-

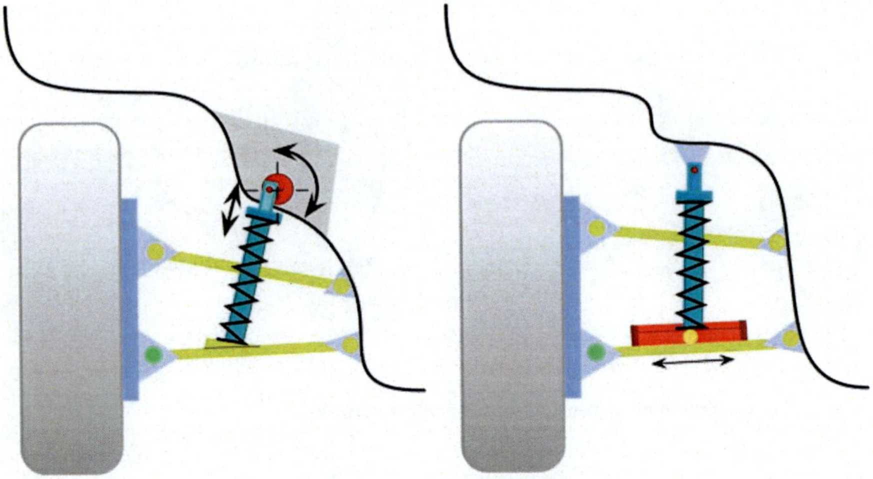

(a) Modification of the upper position of the damper/spring (b) Modification of the lower position of the damper/spring

Fig. 1.4 Concept of series active variable-geometry suspension

tion with smaller complexity. Furthermore, it provides a low increase in unsprung mass and fail-safe operation. An application example of series variable-geometry road suspension for high-speed maneuvers of road vehicles on the rough road can be found in Nazemi et al. (2022). Through the proposed $\mathscr{H}_\infty$ control design method, the performance level of the roll motion can be improved.

In the following, the first solution is detailed, which has been implemented in a test bed (Yu et al. 2019). Figure 1.4a shows that the upper position of the spring-damper unit is modified through the rotation of an electric motor. Thus, the stiffness of the suspension through the modification is significantly influenced. Another solution is to modify the upper position of the damper/spring through a hydraulic cylinder, which is able to provide lateral motion (Anubi et al. 2013).

A further solution is the parallel active variable-geometry suspension, which can be more suitable for heavy vehicles. The purpose of the intervention is similar to the series solution, i.e., to modify suspension stiffness. In this solution, the original suspension construction is not modified, but it is extended with an actuator. It is able to provide a motion for the suspension without the modification of the positions of the joints in the suspension construction. In this concept, a rotary-electromechanical-actuator provides the rotation of a parallel active link, with which the position of the wheel is influenced. Figure 1.5 (Yu et al. 2018) illustrates the solution for double-wishbone suspension, and the works Yu et al. (2021b) and Feng et al. (2022) provide its application to a sport utility vehicle. Nevertheless, it can also be applied for further constructions; see, e.g., Evers et al. (2008b).

A further type of application of the variable-geometry vehicle suspension is its implementation for steering purposes on lightweight narrow vehicles. In this case,

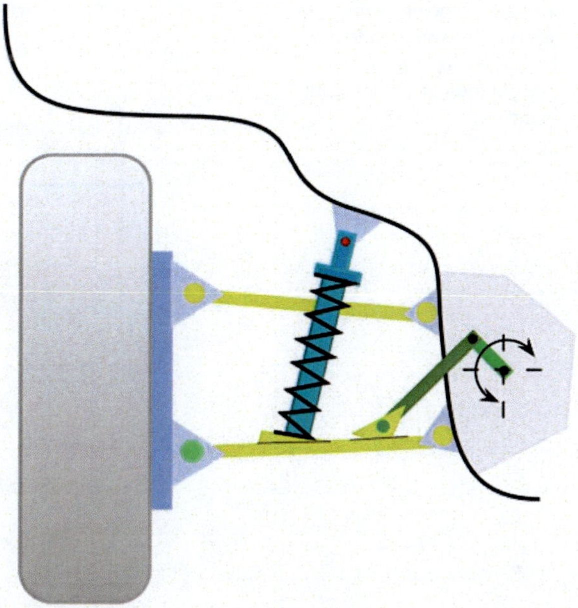

Fig. 1.5 Concept of parallel variable-geometry suspension

the control design on active wheel tilting is requested, with which the steering capability of the vehicle can be improved. This solution is similar to the tilting motion of motorcycles in bends (Suarez 2012). Carrying out precise steering operation with tilting requires an active system, which helps to balance the narrow vehicle for the driver during the cornering maneuver; see Tang and Khajepour (2019). Figure 1.6 shows an exemplary scheme of steering construction from the front view. The cornering maneuver of the vehicle is realized through the tilting of the wheels, which is achieved through the lateral motion of a rod. It can be actuated by the driver or an electric actuator. Since the tilting of the wheels has a significant value, the lateral force in the tire–road contact is highly increased. Moreover, it can result in a high roll angle for the chassis, whose limitation is a challenge in the design of narrow tilting vehicles. Several prototype vehicles have three wheels, in which two wheels can be on the front axle (Piyabongkarn et al. 2004) or on the rear axle (Chiou and Chen 2008).

One of the variable-geometry suspension constructions, whose control design in Part I is presented, can be found in the illustration of Fig. 1.7. In the case of this construction, the lateral motion of a joint is modified, which connects a suspension arm and the chassis; see, e.g., Németh and Gáspár (2012) for McPherson suspension and Németh and Gáspár (2013, 2017), Iman et al. (2010) for double-wishbone suspension. The wheel orientation and position have a direct connection to the lateral vehicle motion, which results in the tire force on the front (Németh and Gáspár 2013) or on the rear wheels (Németh and Gáspár 2011). Although the intervention in this

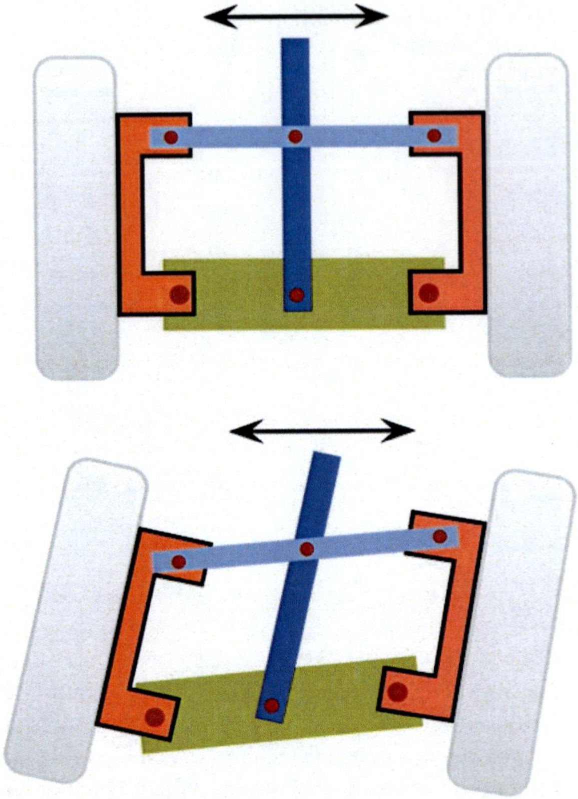

Fig. 1.6 Scheme of the steering construction for narrow tilting vehicles: straight motion and steering maneuver

variable-geometry suspension requires low energy and several functionalities can be achieved, the impact of the camber angle on the lateral wheel force is limited.

The intervention of variable-geometry suspension systems requires the lateral motion of the suspension arm. In a real implementation, it is realized using an electro-hydraulic actuator (Iman et al. 2010; Lee et al. 2006; Schiehlen and Schirle 2006) or an electric motor (Evers et al. 2008a; Arana et al. 2015).

Figure 1.7 shows that upper arm lateral motion results in wheel steering and tilting. The motion can be achieved through a hydraulic cylinder, which requests the positioning of the piston (Németh et al. 2014). The rotation of the crankshaft yields arm lateral motion, which is connected to the cylinder. The control problems of variable-geometry suspension are to provide a lateral arm motion profile, and also to handle hydraulic fault occasions.

The positioning of the actuator and the suspension geometry have a significant impact on the effectiveness of the control intervention. Therefore, designing control and construction are not independent, because both have a high impact on the achieved performance. Since there is a trade-off between the control design and the construction design, an optimization criterion which contained both the performances

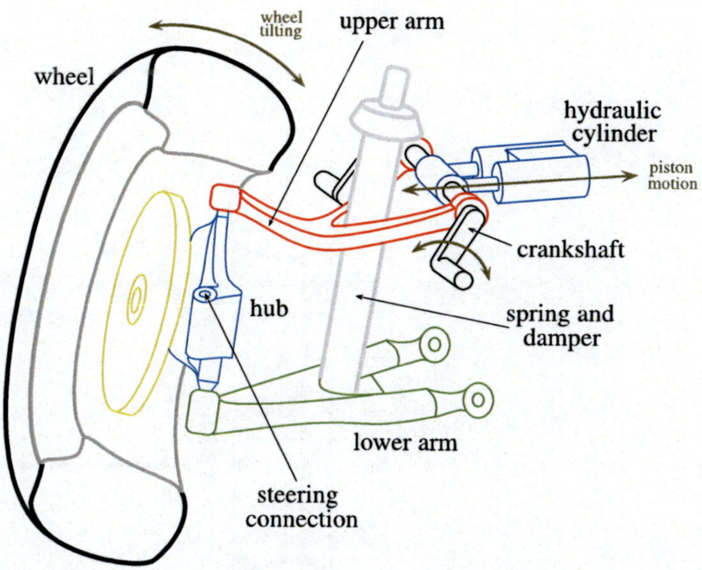

Fig. 1.7 Illustration of the construction of variable-geometry suspension with upper arm lateral motion in the double-wishbone suspension (Reprinted from Németh and Gáspár 2017, with permission from Elsevier)

of the suspension construction and the performances of the control design has been formulated (Németh and Gáspár 2013). The most important performance requirements, such as roll angle minimization, tire wear reduction and path following, can be improved through simultaneous design methods.

Finally, the concept of a variable-geometry suspension is presented, whose operation significantly differs from the previous systems. Figure 1.8 illustrates a McPherson type of suspension on the front axle, with which the steering functionality of the vehicle can be achieved (Fényes et al. 2021; Németh et al. 2019; Wang et al. 2011a; Hu et al. 2015). In this suspension construction, the actuator connects the wheel and wheel hub. Its role is to provide an active torque M_{act} around B, with which the requested camber angle (γ) is achieved. If the camber angle is varied, then the scrub radius r_δ is also influenced. The virtual distance scrub radius can be determined as the radius in the steering mechanism for wheel steering rotating motion. The rotation for the steering is achieved around the virtual axis, which interconnects the connection point A of the chassis and D. Since in the suspension concept the front wheels are driven, the longitudinal forces on the wheels create a moment on the wheels and, thus, realize steering angles through the rotation around the virtual axis. The advantages of this suspension are the integration possibility with the driving and the enhanced independent steering functionality with increased maneuverability. Moreover, the realization of the steering angle requires a low scrub radius, and, thus, the actuator

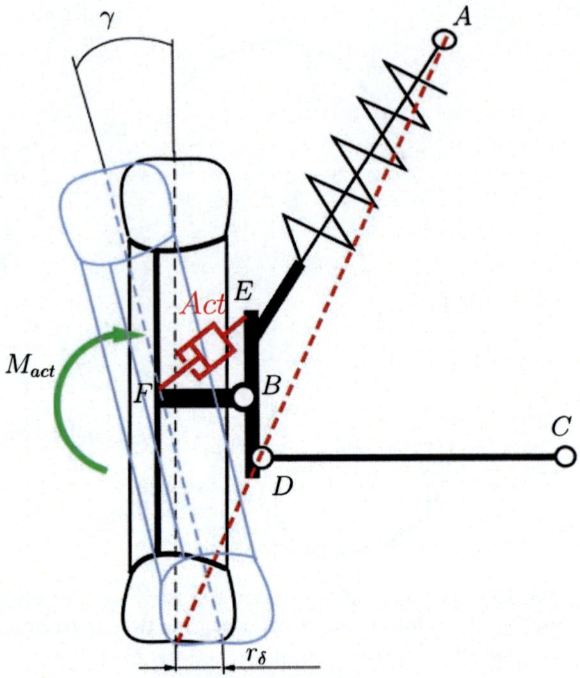

Fig. 1.8 Illustration of variable-geometry-suspension-based on the modification of scrub radius (©2017 IEEE. Reprinted, with permission, from Németh et al. 2017)

with low dynamic capacity can be enough for the intervention. In Part II, the details of the control design and the achievable functionalities and integration possibilities are found.

1.3 Motivation of Using Learning Features in Suspension Control Systems

The growing number of automated functionalities in vehicles, i.e., environmental perception, decision-making, trajectory planning, control and intervention with intelligent actuators, leads to the necessity of designing complex control systems. The complex structure can simultaneously include traditional control systems, such as model-based optimal/robust solutions, and other non-traditional control systems, such as learning-based procedures.

Nonetheless, the growing number of functionalities poses an increasing number of performance requirements against control systems, on which guarantees must be provided. Designing conventional control systems, it is possible to define the system performance in mathematical form. As there may be conflicts among individual performances, a balance must be created between the levels of the performances, e.g., the application of weight selection strategies, and tuning methods. The conventional control design methods have the advantage of ensuring a minimum closed-loop performance level with the obtained control system theoretically. Nevertheless, the formulation of performances for aggregated systems can lead to various difficulties, especially regarding the mathematical description of human–machine interactions; e.g., it may be complicated to formulate human driver intentions, driving characteristics or feelings of traveling comfort. These factors are important from the viewpoint of customer needs, and thus, they in the control design of suspension control must be involved (Savaresi et al. 2010).

Non-conventional control techniques may give a solution regarding the increased number of performance requirements; e.g., using the methods of imitation learning, it is unnecessary to formulate a mathematical description of the performance of the system. The training kit consists of samples on the desired system operation, which meets human requests; see Codevilla et al. (2018) and Wang et al. (2019). Reinforcement learning is also an example of how performance requirements are possible to be maintained through a learning process with a high number of various scenarios; see Dossa et al. (2019) and Chen et al. (2020). However, various types of advanced machine learning techniques can effectively solve different control problems, the resulting level of performance is not theoretically ensured, and, thus, various training samples do not ensure the reduction of performance level under all possible vehicle-traffic scenarios. It poses the challenge of designing control systems with a theoretically proven minimum level of performance guarantees through non-conventional methods. This claims the joint use of traditional, i.e., model-based, and learning-based control approaches. The complication in the synthesis and in the analysis of aggregated systems is yielded by their different mathematical structures, e.g., neural-network-based description and matrix formulation of the dynamic control system.

The literature presents different approaches to ensure the stability and performance guarantees of machine learning-based control systems. Åkesson and Toivonen (2006) and Zhang et al. (2016) present the application of the machine learning method together with Model Predictive Control (MPC) techniques. Since computation time can be a critical issue in the online computation of the MPC problem, the benefit of using learning-based solutions is to improve the real-time operation of the system; e.g., through neural networks (NN), the control input of MPC through the learning of the optimal solution in different scenarios can be approximated. A disadvantage of this method is that it is difficult to ensure constraints and system stability, as described in Hertneck et al. (2018). Karg and Lucia (2019) has developed a statistical verification strategy and McKinnon and Schoellig (2019) has realized a stochastic MPC method for providing guarantees. Koller et al. (2018) has applied a recursive design method with a constraint on the terminal set, with which safe control actions at each iteration step have been ensured. A repetitive learning solution in the framework of MPC design can be found in Rosolia and Borrelli (2018), in which terminal sets and costs in a recursive way have been constructed. The iterative construction process is based on the enlargement of the terminal properties using the pairs of state-input trajectories, which guarantees the non-decreasing property of performance requirements and feasibility of the control task. This procedure offers an appropriate method of providing guarantees, since the functionality of learning is integrated into the model predictive framework, e.g., learning the terminal cost and terminal set via the iteration process. Nevertheless, the achievements of deep learning methods in this MPC design cannot be integrated. Set-based approach to achieve the state trajectories keeping in safe sets during the training process is provided by Larsen et al. (2017). Hamilton–Jacobi reachability methods can be used to provide a safety framework, in which different types of learning techniques can be involved; see, e.g., the work of Fisac et al. (2019).

In other techniques, the roles in control tasks of different control elements, i.e., model-based controllers and learning-based controllers, are separated, but meanwhile they function together. As an example, Zhai et al. (2007) suggests a method, in which the linear control guarantees stability for the system and the adaptive neural network (ANN) with state and output feedback deals with the handling of nonlinear characteristics of the system. The method is applied for nonaffine discrete-time uncertain systems with one input and one output. Likewise, a state-of-the-art switching controller has been introduced in Zhao et al. (2016), which contains a conventional ANN and moreover, an additional robust controller, which retracts transients from outside of the approximation range. The contribution of the method is that the output converges into a bounded range of the reference signal, which guarantees the global stability of the closed loop. A further solution to the control problem is to design model-reference adaptive control, which provides guarantees on tracking-based performance requirements, even if dynamic uncertainties in the system are found (Calliess 2019).

Several other research focus on guarantees for reinforcement-learning-based methods (RL), for which methods providing performance guarantees is challenging. Pecka and Svoboda (2014) presents a review of methods for safe exploration with various safety definitions in the context of RL-based controllers. The work of Mao et al. (2019) provides a safe RL-based method, in which the model for learning is trained online, and if an unsafe state-space region is reached, safety through a backup policy is restored. Mannucci et al. (2018) presents a control design approach that is based on interval estimation of the dynamics of the agent, while it is assumed that the agent has limited ability to detect the presence of incoming fatal states. The combination of RL and robust control design is presented in Kretchmar et al. (2001). In this process, a robust control system and a NN in a parallel structure are ordered. During the training process, the performance level of the closed-loop system is improved. Stability of the system by considering the NN as an uncertain system from the viewpoint of robust control is achieved. Cheng et al. (2019) proposed a control scheme, which integrates a model-based conventional controller and model-free-structured RL control. In this solution control barrier functions are used, which provides the possibility of learning unknown system dynamics online to guarantee safety. Lyapunov-based techniques and statistical approaches are applied for the verification of guarantees on system stability by Berkenkamp et al. (2017). Analysis of the smoothness margin of NNs, which are trained through the RL process, is introduced by Jin and Lavaei

(2018). The examination is formed on a semi-definite programming-based feasibility task, which does not require information on the trained controller parameters.

This brief survey reveals that a systematic global solution for providing guarantees on performance requirements with learning-based controllers in the current literature is not found. Integration of the learning-based methods by iterations and agent analysis into control design has limitations. Nevertheless, training of learning-based controls can have efficient methods for different problems, which have been well-developed in recent years. Thus, it can be a fruitful way to make them part of the control system design without substantial changes. These goals have motivated the research for enhanced control of variable-geometry vehicle suspension in this book, which has been supported by the European Union within the framework of the National Laboratory for Autonomous Systems (RRF-2.3.1-21-2022-00002) and the National Research, Development and Innovation Office (NKFIH) under OTKA Grant Agreement No. K 135512 "Robust Control Design for Automated Vehicles with Guaranteed Performances".

With the motivation of automated vehicle systems, particularly suspension systems, the focus of the present book is on the application of learning-based methods on the given types of subsystems, such as variable-geometry vehicle suspensions, and active and semi-active suspensions. A substantial topic in the field of semi-active suspension systems is the allocation of the actuation with automated vehicle functionalities, e.g., automated speed selection provides energy-efficient, safe and comfortable movement, as described in Du et al. (2018). The requirements on safety, energy-based performance and traveling comfort against road roughness can also be ensured as a part of control on vertical dynamics. Its reason is that choosing speed and route profiles may not be handled independently from the control design of suspension without loss of performance level. Therefore, it is recommended to involve expected road information in the design of the suspension control, such as through preview control (Göhrle et al. 2013; Caliskan et al. 2016) or road estimation (Nguyen et al. 2016). A disadvantage of the solution is that enhanced handling of the forthcoming road can request different sources of information, and moreover, an extensive consideration of vehicle motions, for which non-standard control solutions are necessary to apply; e.g., LiDAR signals must be processed through deep learning techniques to detect road unevenness along the forthcoming road sections; see Alcantara et al. (2016), Kim et al. (2018) and Ming et al. (2020). However, there are a number of solutions for the control design of semi-active suspension systems (for example, Skyhook Do et al. 2010, $\mathcal{H}_\infty$, gain scheduling Savaresi et al. 2010 and predictive methods Nguyen et al. 2016); the introduced control structure has the benefit of the ability to perform preview and the ability to use an increased amount of external signals.

Accordingly, the book contributes a design framework for variable-geometry vehicle suspension control design that improves lateral and vertical performance levels. This is reached by using road information ahead of the vehicle along a limited horizon, which through a NN in the control system is involved. In the suggested design method, the nonlinearities of the system dynamics are developed by robust control design. The resulting control scheme includes LPV-based and non-conventional controllers with different roles in the intervention of the active suspension system. The role of the non-conventional controller is to improve the performance level of lateral vehicle motion, with which it can be close to the style of a human driver. Nevertheless, this agent must be supervised by a conventional controller to achieve limitations on the path following error. As a result, the control system provides safe motion of the vehicle, and the variable-geometry suspension control enforces comfortable driving, too.

1.4 Contents of the Book

Concerning variable-geometry suspension systems, the book is divided into three main sections, i.e., different modeling and control design methods for wheel tilting and independent steering, and the applicability of learning methods. The motivation of the book is presented in this chapter. It also presents constructions, modeling and control design methods proposed in the literature.

The scope of Part I is to provide modeling and control methods for the wheel tilting of automated vehicles. The advantage of wheel tilting control is to meet the predefined performance specifications, such as the enhanced maneuvering capability of the vehicle, the reduction of tire wear and the minimization of chassis roll angle. The chapters of Part I propose modeling and control design methods to achieve appropriate wheel tilting.

Chapter 2 presents mathematical models to describe the kinematics and dynamics of variable-geometry suspensions. Analytic and iterative solutions on the models are provided, and the transformation of complex suspension models to simplified control-oriented models is presented. Here, the model is formed as an Linear Parameter-Varying (LPV) structure in order to prepare the standard LPV control design and achieve the predefined performance specification. Chapter 3 presents the LPV-based control design based on the control-oriented suspension model. This chapter proposes an iterative method, with which the control system and some constructional parameters of the suspension in a joint process can be designed. Moreover, the design process on the level of the electro-hydraulic actuator is also presented, in which the parameter variations of the hydraulic system are involved. Chapter 4 presents the modeling, analysis and control of variable-geometry-suspension-based on a nonlinear polynomial approach, thus the Sum-of-Squares (SOS) method is introduced. The nonlinear analysis provides insight into the lateral dynamics of the system. The results of the analysis in the control design procedure are incorporated. The operation of the method is presented through a software-in-the-loop environment. From the hardware side of the simulation running DSpace Autobox is used, and from the software side high-fidelity CarSim simulator is applied.

The scope of Part II is to present independent steering with variable-geometry suspension. This solution has a significant impact on automated vehicles, especially on lightweight urban electric vehicles. The chapters of Part II provide mathematical background for modeling and control design, the coordination method of torque vectoring functionality and independent steering on the front wheels, and, finally, the implementation of the control system on a suspension test bed is demonstrated.

Chapter 5 presents modeling methods for the dynamics of electro-hydraulic actuators, together with variable-geometry vehicle suspension dynamics and lateral vehicle dynamics. Here, complex nonlinear models are formed, which are used to derive control-oriented models. Chapter 6 presents a complex control design based on a hierarchical structure for the variable-geometry suspension. This hierarchical structure contains different levels for lateral control, wheel steering control and suspension control. This chapter also provides design methods on each level, in which LPV control design methods are applied. Chapter 7 focuses on the coordinated control design of independent steering and torque vectoring functions. The coordination is based on a reconfiguration strategy, in which the capability of control interventions is examined with an SOS-based analysis. In this chapter, the formulation of SOS programming for the given task and the reconfiguration strategy are proposed. In Chap. 8, the operation of the control design for independent steering is shown in various test scenarios on a test bed with hardware-in-the-loop simulations. This chapter presents the construction, the control methods and a few illustrative examples.

The emerging trend of autonomous vehicles, i.e., the integration of control design and learning methods to achieve enhanced suspension functions, is the focus of Part III. The current task of the integration is to provide methods with which the performance specifications of the controlled system are achieved.

Chapter 9 focuses on data-driven modeling and control design methods for variable-geometry suspension. The dataset comes from measurements on the suspension test bed. In this chapter, the extension of the LPV-based modeling and control methods with the data-driven method is presented. Chapter 10 presents the theoretical background of variable-geometry suspension control, where in the closed loop a learning-based agent is incorporated. Closed-loop system robustness can be achieved by a LPV design extended with an iterative procedure. In Chap. 11, the application of the iterative learning method applied for the control design of variable-geometry suspension is shown. The purpose of the method is to guarantee the limitation of lateral error of trajectory tracking, in which independent steering is applied. Two application examples are presented. The first example presents the operation of a NN-based agent in a control, which results in supervised learning. The second example shows driver-in-the-loop scenarios, in which performances are guaranteed.

References

Åkesson BM, Toivonen HT (2006) A neural network model predictive controller. J Proc Control 16(9):937–946

Alcantara DH, Morales-Menendez R, Mendoza RAR (2016) Teaching semi-active suspension control using an experimental platform. In: 2016 American control conference (ACC), pp 7334–7339

Anubi OM, Patel DR, Crane CD III (2013) A new variable stiffness suspension system: passive case. Mech Sci 4(1):139–151

Arana C, Evangelou SA, Dini D (2015) Series active variable geometry suspension for road vehicles. IEEE/ASME Trans Mechatron 20(1):361–372

Berkenkamp F, Turchetta M, Schoellig A, Krause A (2017) Safe model-based reinforcement learning with stability guarantees. In: Guyon I, Luxburg UV, Bengio S, Wallach H, Fergus R, Vishwanathan S, Garnett R (eds) Advances in neural information processing systems 30, Curran Associates, Inc., pp 908–918

Braghin F, Cimatti F, Sabbioni E (2008) Development of a variable kinematics suspension system. In: 11th mini conference on vehicle system dynamics, identification and anomalies, Budapest pp 1–6

Caliskan K, Henze R, Kukukay F (2016) Potential of road preview for suspension control under transient road inputs. IFAC-PapersOnLine 49(3):117 – 122, 14th IFAC Symposium on Control in Transportation Systems

Calliess J (2019) Online optimisation for online learning and control - from no-regret to generalised error convergence. In: $18th$ European control conference (ECC), pp 2480–2485

Castro R, Araújo RE, Tanelli M, Savaresi SM, Freitas D (2012) Torque blending and wheel slip control in evs with in-wheel motors. Veh Syst Dyn 50:71–94

Cheng R, Orosz G, Murray R, Burdick J (2019) End-to-end safe reinforcement learning through barrier functions for safety-critical continuous control tasks. In: The Thirty-Third AAAI conference on artificial intelligence, pp 3387–3395

Chen S, Wang M, Song W, Yang Y, Li Y, Fu M (2020) Stabilization approaches for reinforcement learning-based end-to-end autonomous driving. IEEE Trans Veh Technol 1–1

Chiou J, Chen C (2008) Modeling and verification of a diamond-shape narrow-tilting vehicle. IEEE/ASME Trans Mechatron 13(6):678–691

Codevilla F, Muller M, Lopez A, Koltun V, Dosovitskiy A (2018) End-to-end driving via conditional imitation learning. In: 2018 IEEE international conference on robotics and automation (ICRA), pp 4693–4700

Do A, Sename O, Dugard L, Savaresi S, Spelta C, Delvecchio D (2010) An extension of mixed sky-hook and add to magneto-rheological dampers. IFAC Proc Vol 43(21):25–31, 4th IFAC symposium on system structure and control

Doericht V (2013) Strategic visioning - future of business. In: Möhrle MG, Isenmann R, Phaal R (eds) Technology roadmapping for strategy and innovation. Springer, Charting the Route to Success

Dossa RFJ, Lian X, Nomoto H, Matsubara T, Uehara K (2019) A human-like agent based on a hybrid of reinforcement and imitation learning. In: 2019 international joint conference on neural networks (IJCNN), pp 1–8

Du Y, Liu C, Li Y (2018) Velocity control strategies to improve automated vehicle driving comfort. IEEE Intell Transp Syst Mag 10(1):8–18

EUCAR (1992) The automotive industry R and D challenges of future. In: European council for automotive R&D (EUCAR). http://www.eucar.be/publications/

Evers W, van der Knaap A, Besselink I, Nijmeijer H (2008a) Analysis of a variable geometry active suspension. In: International symposium on advanced vehicle control, Kobe, Japan, pp 1–6

Evers W, van der Knaap A, Besselink I, Nijmeijer H (2008b) Modeling, analysis and control of a variable geometry actuator. In: Intelligent vehicles symposium, Eindhoven, Netherlands

Fallah MS, Bhat R, Xie WF (2009) New model and simulation of Macpherson suspension system for ride control applications. Veh Syst Dyn 47(2):195–220

Feng Z, Yu M, Evangelou SA, Jaimoukha IM, Dini D (2022) Mu-synthesis pid control of full-car with parallel active link suspension under variable payload 2203.04147

Fényes D, Németh B, Gáspár P (2018) Control design of variable-geometry suspension systems using a reconfiguration strategy. In: 2018 IEEE 18th international symposium on computational intelligence and informatics (CINTI), pp 000,081–000,086

Fényes D, Németh B, Gáspár P (2021) Data-driven modeling and control design in a hierarchical structure for a variable-geometry suspension test bed. In: 2021 60th IEEE conference on decision and control (CDC), pp 5047–5052

Fisac JF, Akametalu AK, Zeilinger MN, Kaynama S, Gillula J, Tomlin CJ (2019) A general safety framework for learning-based control in uncertain robotic systems. IEEE Trans Autom Control 64(7):2737–2752

Gáspár P, Szabó Z, Bokor J, Németh B (2017) Robust control design for active driver assistance systems. A linear-parameter-varying approach. Springer

Göhrle C, Schindler A, Wagner A, Sawodny O (2013) Model predictive control of semi-active and active suspension systems with available road preview. In: 2013 European control conference (ECC), pp 1499–1504

Goodarzia A, Oloomia E, Esmailzadehb E (2010) Design and analysis of an intelligent controller for active geometry suspension systems. Veh Syst Dyn 49(1):333–359

Gough V, Shearer G (1956) Front suspension and tyre wear. In: Proceedings of the automobile division, The Institution of Mechanical Engineers, pp 171–216

Habibi H, Shirazi KH, Shishesaz M (2008) Roll steer minimization of McPherson-strut suspension system using genetic algorithm method. Mech Mach Theory 43:57–67

Heissing B, Ersoy M (2011) Chassis handbook, fundamentals, driving dynamics, components, mechatronics, perspectives. Vieweg-Teubner Verlag, ATZ/MTZ-Fachbuch

Hertneck M, Köhler J, Trimpe S, Allgöwer F (2018) Learning an approximate model predictive controller with guarantees. IEEE Control Sys Lett 2(3):543–548

Hong KS, Jeon DS, Sohn HC (1999) A new modeling of MacPherson suspension system and its optimal pole-placement control. In: 7th IEEE mediterran control conference on control and automation, pp 559–579

Hu JS, Yin D, Hori Y (2011) Fault-tolerant traction control of electric vehicles. Control Eng Pract 204–213

Hu C, Jing H, Wang R, Yan F, Chen N (2015) Fault-tolerant control of fwia electric ground vehicles with differential drive assisted steering. IFAC-PapersOnLine (9th IFAC symposium on fault detection, supervision and safety for technical processes, Paris, France) 48(21):1180–1185

Iman M, Esfahani M, Mosayebi M (2010) Optimization of double wishbone suspension system with variable camber angle by hydraulic mechanism. Eng Technol 299–306

Jin M, Lavaei J (2018) Control-theoretic analysis of smoothness for stability-certified reinforcement learning. In: 2018 IEEE conference on decision and control (CDC), pp 6840–6847

Karg B, Lucia S (2019) Learning-based approximation of robust nonlinear predictive control with state estimation applied to a towing kite. In: $18th$ European control conference (ECC), pp 16–22

Kim S, Kim H, Kang D (2018) Vibration control of a vehicle active suspension system using a ddpg algorithm. In: 2018 18th international conference on control, automation and systems (ICCAS), pp 1654–1656

Koller T, Berkenkamp F, Turchetta M, Krause A (2018) Learning-based model predictive control for safe exploration. In: 2018 IEEE conference on decision and control (CDC), pp 6059–6066

Kretchmar RM, Young PM, Anderson CW, Hittle DC, Anderson ML, Delnero CC (2001) Robust reinforcement learning control with static and dynamic stability. Int J Robust Nonlinear Control 11(15):1469–1500

Larsen RB, Carron A, Zeilinger MN (2017) Safe learning for distributed systems with bounded uncertainties. IFAC-PapersOnLine 50(1):2536–2542, 20th IFAC world congress

Lee HG, Won CJ, Kim JW (2009) Design sensitivity analysis and optimization of mcpherson suspension systems. World Congr Eng 1532–1537

Lee U (2010) Active geometry control suspension. ATZ Worldwide 4–9

Lee UK, Han CS (2001) A suspension system with a variable roll centre for the improvement of vehicle handling characteristics. Proc IMechE Part D: J Autom Eng 215(6):677–696

Lee H, Lee U, Ha S, Han C (1999) Four-wheel independent steering (4wis) system for vehicle handling improvement by active rear toe control. JSME Int J Ser C 42(4):947–956

Lee S, Sung H, Kim J, Lee U (2006) Enhancement of vehicle stability by active geometry control suspension system. Int J Autom Technol 7(3):303–307

Lee UK, Lee SH, Han CS, Hedrick K, Catala A (2008) Active geometry control suspension system for the enhancement of vehicle stability. Proc IMechE Part D: J Autom Eng 222(6):979–988

Lee S, Sung H, Lee U (2005) A study to the enhancement of vehicle stability by Active Geometry Control Suspension (AGCS) system. In: 13th international pacific conference on automotive engineering, Gyeongju, Korea, pp 1–6

Lin C, Xu Z (2015) Wheel torque distribution of four-wheel-drive electric vehicles based on multi-objective optimization. Energies 8(5):3815–3831

Lugner P (2019) Vehicle dynamics of modern passenger cars. CISM international centre for mechanical sciences. Springer International Publishing

Malvezzi F, Hess Coelho T, Orsino R (2022) Feasibility and performance analyses for an active geometry control suspension system for over-actuated vehicles. J Braz Soc Mech Sci Eng 44(178)

Mannucci T, van Kampen E, de Visser C, Chu Q (2018) Safe exploration algorithms for reinforcement learning controllers. IEEE Trans Neural Netw Learn Syst 29(4):1069–1081

Mao H, Schwarzkopf M, He H, Alizadeh MI (2019) Towards safe online reinforcement learning in computer systems. In: 33rd conference on neural information processing systems, Vancouver, Canada, December 8

Martins I, Esteves J, Marques GD, Pina da Silva F (2006) Permanent-magnets linear actuators applicability in automobile active suspensions. IEEE Trans Veh Technol 55(1):86–94

McKinnon CD, Schoellig AP (2019) Learning probabilistic models for safe predictive control in unknown environments. In: $18th$ European control conference (ECC), pp 2472–2479

Ming L, Yibin L, Xuewen R, Shuaishuai Z, Yanfang Y (2020) Semi-active suspension control based on deep reinforcement learning. IEEE Access 8:9978–9986

Nazemi S, Masih-Tehrani M, Mollajafari M (2022) GA tuned hinf roll acceleration controller based on series active variable geometry suspension on rough roads. Int J Veh Perfor 8(2–3):166–187

Németh B, Gáspár P (2012) Mechanical analysis and control design of McPherson suspension. Int J Veh Syst Model Test 7(2):173–193

Németh B, Gáspár P (2013) Control design of variable-geometry suspension considering the construction system. IEEE Trans Veh Technol 62(8):4104–4109

Németh B, Gáspár P (2017) Nonlinear analysis and control of a variable-geometry suspension system. Control Eng Pract 61(April):279–291

Németh B, Fényes D, Gáspár P, Bokor J (2019) Coordination of independent steering and torque vectoring in a variable-geometry suspension system. IEEE Trans Control Syst Technol 27(5):2209–2220

Németh B, Fényes D, Gáspár P, Bokor J (2017) Control design of an electro-hydraulic actuator for variable-geometry suspension systems. In: 2017 25th Mediterranean conference on control and automation (MED), pp 180–185

Németh B, Gáspár P (2011) Integration of control design and variable geometry suspension construction for vehicle stability enhancement. In: 2011 50th IEEE conference on decision and control and european control conference, pp 7452–7457

Németh B, Varga B, Gáspar P (2014) Design of a variable-geometry suspension system to enhance road stability. In: 2014 22nd mediterranean conference of control and automation (MED), pp 55–60

Nguyen MQ, Canale M, Sename O, Dugard L (2016) A model predictive control approach for semi-active suspension control problem of a full car. In: 2016 IEEE 55th conference on decision and control (CDC), pp 721–726

Pecka M, Svoboda T (2014) Modelling and simulation for autonomous systems, Lecture notes in computer science, vol 8906. Springer, Cham. Chap Safe exploration techniques for reinforcement learning - an overview

Pernebo L, Hansson B (2002) Plug and play in control loop design. Reglermöte, Linköping, Sweden

Piyabongkarn D, Keviczky T, Rajamani R (2004) Active direct tilt control for stability enhancement of a narrow commuter vehicle. Int J Autom Technol 5(2):77–88

Rajaie MA, Khajepour A, Pazooki A, Soltani A (2019) Development of a novel integrated corner module for narrow urban vehicles. Proc Inst Mech Eng Part D: J Autom Eng 233(3):548–556

Ronci M, Artuso P, Bocci E (2011) Four independent wheels steering system analysis. SAE Technical Paper. https://doi.org/10.4271/2011-01-0241

Rosolia U, Borrelli F (2018) Learning model predictive control for iterative tasks. a data-driven control framework. IEEE Transactions on Automatic Control 63(7):1883–1896

Sancibrian R, Garcia P, Viadero F, Fernandez A, De-Juan A (2010) Kinematic design of double-wishbone suspension systems using a multiobjective optimisation approach. Veh Syst Dyn 48(7):793–813

Savaresi S, Poussot-Vassal C, Spelta C, Sename O, Dugard L (2010) Semi-active suspension control design for vehicles. Butterworth-Heinemann

Schiehlen W, Schirle T (2006) Modeling and simulation of hydraulic components for passenger cars. Veh Syst Dyn: Int J Veh Mech Mobil 44(sup1):581–589

Sharp R (1998) Variable geometry active suspension for cars. IEEE Comput Control Eng J 9(5):217–222

Shuai Z, Zhang H, Wang J, Li J, Ouyang M (2013) Lateral motion control for four-wheel-independent-drive electric vehicles using optimal torque allocation and dynamic message priority scheduling. Control Eng Pract 24:55–66

Stoustrup J (2009) Plug & play control: control technology towards new challenges. Eur J Control 15(3):311–330

Suarez L (2012) Active tilt and steer control for a narrow tilting vehicle: control design and implementation. Lambert Academic Publishing

Tang C, Khajepour A (2019) Narrow tilting vehicles: mechanism, dynamics, and control. Springer

Vukobratovic M, Potkonjak V (1999) Modeling and control of active systems with variable geometry. I: general approach and its application. Mech Mach Theory 35:179–195

Wang J, Wang Q, Jin L, Song C (2011a) Independent wheel torque control of 4WD electric vehicle for differential drive assisted steering. Mechatronics 21:63–76

Wang R, Chen Y, Feng D, Huang X, Wang J (2011b) Development and performance characterization of an electric ground vehicle with independently actuated in-wheel motors. J Power Sources 196(8):3962–3971

Wang Q, Chen L, Tian B, Tian W, Li L, Cao D (2019) End-to-end autonomous driving: an angle branched network approach. IEEE Trans Veh Technol 68(12):11,599–11,610

Wang R, Wang J (2012) Fault-tolerant control for electric ground vehicles with independently-actuated in-wheel motors. J Dyn Syst Measur Control 134

Watanabe Y, Sharp RS (1999) Mechanical and control design of a variable geometry active suspension system. Veh Syst Dyn 32(2):217–235

Wu FK, Yeh T, Huang CF (2013) Motor control and torque coordination of an electric vehicle actuated by two in-wheel motors. Mechatronics 23:46–60

Yu M, Arana C, Evangelou SA, Dini D, Cleaver GD (2018) Parallel active link suspension: a quarter-car experimental study. IEEE/ASME Trans Mechatron 23(5):2066–2077

Yu M, Arana C, Evangelou SA, Dini D (2019) Quarter-car experimental study for series active variable geometry suspension. IEEE Trans Control Syst Technol 27(2):743–759

Yu M, Cheng C, Evangelou SA, Dini D (2021a) Series active variable geometry suspension: full-car prototyping and road testing. IEEE/ASME Trans Mechatron

Yu M, Evangelou SA, Dini D (2021b) Parallel active link suspension: full car application with frequency-dependent multiobjective control strategies. IEEE Trans Control Syst Technol 1–16

Zhai L, Chai T, Ge SS (2007) Stable adaptive neural network control of nonaffine nonlinear discrete-time systems and application. 2007 IEEE 22nd International Symposium on Intelligent Control 602–607

Zhang T, Kahn G, Levine S, Abbeel P (2016) Learning deep control policies for autonomous aerial vehicles with mpc-guided policy search. In: 2016 IEEE international conference on robotics and automation (ICRA), pp 528–535

Zhao D, Chen W, Wu J, Li J (2016) Globally stable adaptive tracking control for uncertain strict-feedback systems based on neural network approximation. Asian J Control 18(2):527–538

Part I
Variable-Geometry Suspension for Wheel Tilting Control

Chapter 2
LPV-Based Modeling of Variable-Geometry Suspension

This chapter proposes model formulation and analysis methods for variable-geometry suspension systems. Thus, the formulation of enhanced methods is proposed, which are based on the extensions of existing lateral vehicle models, and, furthermore, new methods for modeling and analysis are also provided.

2.1 Lateral Vehicle Model Extension with Wheel Tilting Effect

The main goal of this section is to model the dynamic effect of the proposed variable-geometry suspension. As an example, Fig. 2.1 illustrates the construction of a double-wishbone variable-geometry vehicle suspension. In the case of this construction, the upper arm is moved to the lateral direction (see points A_1 and A_2), which is the control intervention of the system. This motion results in the modification of wheel orientation and position, i.e., suspension geometry is changed by altering the camber angle, which affects the rotation of the front wheel. Moreover, camber angle γ can change the rotation of the wheel around an axis in the case of double-wishbone suspensions. It results in the modification of the steering, which is influenced by the positions of points K and D, i.e., the connection of the steering track-rod and its attachment to the lower arm. Therefore, it is possible to improve the lateral force of the tire with both the angle of steering and the angle of the camber. Consequently, the lateral tire forces can be influenced by the variable-geometry suspension through these signals. The next two sections examine the dynamic effects on steering and wheel tilt by variable-geometry suspension. For this purpose, the lateral forces derived from wheel camber angle modification must be defined primarily.

Although the Magic formula (Pacejka 2004) can provide a detailed description of the lateral force, for numerical reasons and control-oriented modeling, a simplified form is prepared. The lateral tire forces linearly approach the lateral slip angles of the tire in the direction of wheel contact with the ground in the case of small lateral

B. Németh and P. Gáspár, *Control of Variable-Geometry Vehicle Suspensions*, Advances in Industrial Control, https://doi.org/10.1007/978-3-031-30537-5_2

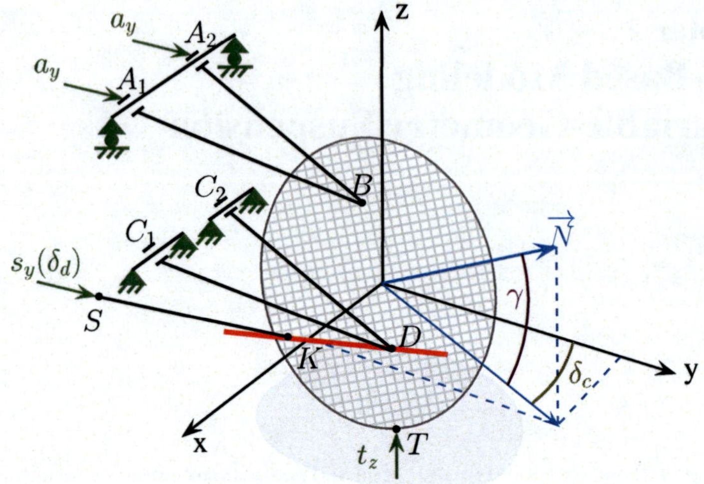

Fig. 2.1 Exemplary scheme of the double-wishbone suspension

slips α_f, α_r. By modifying the vertical offset, the impact of γ can be extended to the linearized tire model, as depicted with different lines in Fig. 2.2. The resulting tire model for control design purposes approximates the lateral force on the wheel as (Németh and Gáspár 2011a)

$$F_i = C_i\delta + C_{i,\gamma}\gamma, \tag{2.1}$$

where C_i is cornering stiffness, $C_{i,\gamma}$ is tilting stiffness, whose coefficient characterizes the value of the offset. Then, the two-wheel vehicle model has been augmented with (2.1) to account for the effect of wheel camber as follows:

$$J\ddot{\psi} = C_1 l_1\left(\delta_d + \delta_c - \beta - \frac{l_1\dot{\psi}}{v}\right) - C_2 l_2\left(-\beta + \frac{l_1\dot{\psi}}{v}\right) + C_{1,\gamma} l_1\gamma, \tag{2.2}$$

$$mv(\dot{\psi} + \dot{\beta}) = C_1\left(\delta_d + \delta_c - \beta - \frac{l_1\dot{\psi}}{v}\right) + C_2\left(-\beta + \frac{l_1\dot{\psi}}{v}\right) + C_{1,\gamma}\gamma, \tag{2.3}$$

where steering angle δ is separated into two parts: $\delta_d + \delta_c$. Here, δ_d stands for the driver steering angle and δ_c is the steering angle affected by a suspension with variable geometry. The δ_c and γ control signals are not independent, both of them depend on the construction and the operation of the variable-geometry suspension.

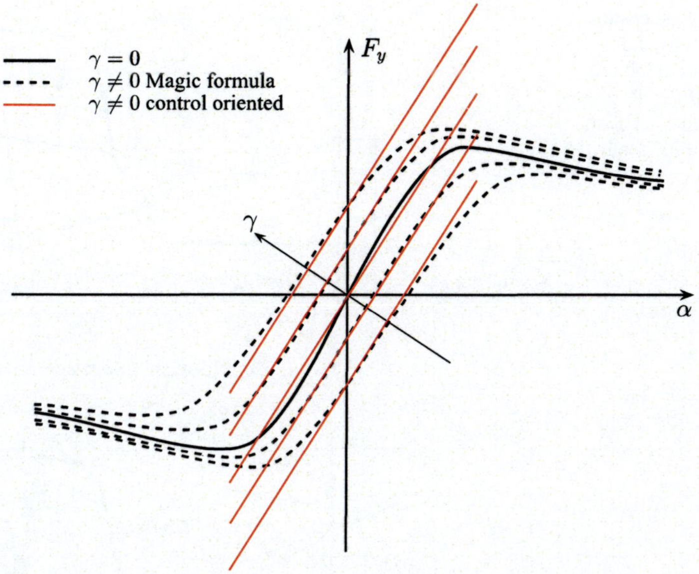

Fig. 2.2 Characteristics of lateral force depending on wheel tilting (©2012 IEEE. Reprinted, with permission, from Németh and Gáspár 2011b)

2.2 Model Formulation of Variable-Geometry Vehicle Suspensions

2.2.1 Formulation of Suspension Kinematics

This subsection details the formulation of variable-geometry mechanism kinematics for two different constructions, i.e., double-wishbone type of suspension (Fig. 2.3a) and McPherson construction (Fig. 2.3b). The description of kinematics requires a geometrical model of the suspensions, which is illustrated in Fig. 2.3c, whose role is to provide calculation on the positions of the suspension arms and the wheel. In this model, mass, inertia and flexibility of the structural elements are not considered, i.e., suspension arms are modeled as bar elements. The motion of the points in a coordinate system is described, which is attached to the chassis of the vehicle. Therefore, chassis roll and road distortions have the same effect on wheel movement.

Tilting of the wheel is determined by the geometric positions of the suspension points, which are influenced by the control input of the mechanism and road distortions. Vertical forces (e.g., due to loads) in the suspension are considered indirectly in modeling the motion of the suspension (Németh and Gáspár 2011c, 2012b). The effect of chassis movement is similar to road distortions. The conversion of the double-wishbone and McPherson suspension including all parameters of a quarter-car model is described in Reimpell (1976).

Fig. 2.3 Suspension constructions and mathematical models (©2012 IEEE. Reprinted, with permission, from Németh and Gáspár 2011b)

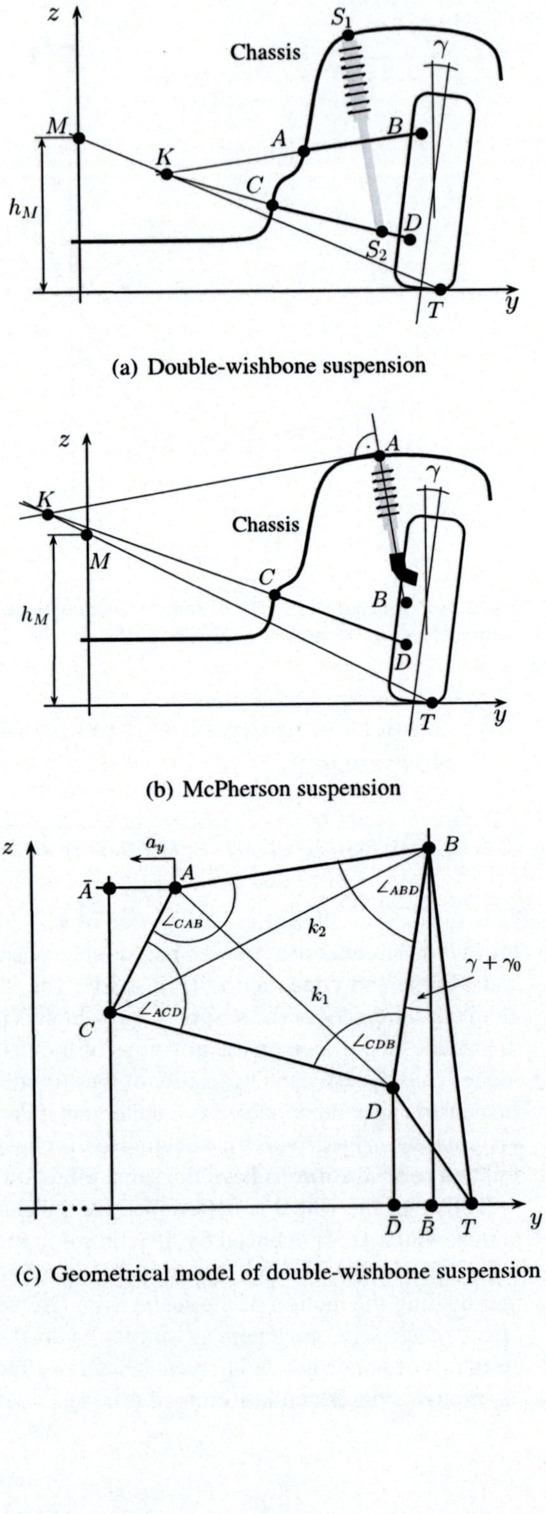

In the case of the presented suspension structures, the control inputs are different. For double-wishbone suspension, the point A of a variable-geometry suspension can only move in a horizontal direction. The real input of the mechanism is the change of point A in the y direction which is designated by a_y. In the case of McPherson suspension, the intervention of the suspension is c_y, which is the lateral motion of C. The aim of the model formulation is to form the relationships between a_y, γ and c_y, γ. Since wheel tilting is also influenced by road roughness, i.e., t_y, t_z motions, their impact on the formulation must also be considered. It can also require the description of point motions B and D.

2.2.2 *Analytic Solution on the Motion of Double-Wishbone Suspension*

The model for motion description of the double-wishbone type of suspension is illustrated in Fig. 2.3c. During the analysis of variable-geometry suspension, the following angles of rectangle $ABCD$ are applied: $\angle_{CAB} + \angle_{ABD} + \angle_{ACD} + \angle_{CDB} = 360°$. Rectangle sides L_{AB}, L_{BD}, L_{CD} have constant length values, depending on the construction, but the length of L_{AC} is changeable. L_{AC} can be calculated from the other sides:

$$L_{AC} = \sqrt{(A_z - C_z)^2 + (A_y - a_y - C_y)^2}. \tag{2.4}$$

In addition, length L_{AC} is determined by the control intervention a_y. Description of the diagonals of a rectangle according to the law of cosine is as follows:

$$k_1^2 = L_{AC}^2 + L_{CD}^2 - 2L_{AC}L_{CD}\cos\angle_{ACD} = L_{BD}^2 + L_{AB}^2 - 2L_{BD}L_{AB}\cos\angle_{ABD}, \tag{2.5}$$

$$k_2^2 = L_{CD}^2 + L_{BD}^2 - 2L_{CD}L_{BD}\cos\angle_{CDB} = L_{AC}^2 + L_{AB}^2 - 2L_{AC}L_{AB}\cos\angle_{CAB}. \tag{2.6}$$

During the examination, the value $\angle_{ACD}$ is chosen as a parameter, from which the angles of the rectangle are calculated. Value of $\angle_{ABD}$ can be calculated using the formula (2.5), such as

$$\angle_{ABD} = \arccos\frac{2L_{CD}L_{AC}\cos\angle_{ACD} + L_{BD}^2 + L_{AB}^2 - L_{AC}^2 - L_{CD}^2}{2L_{AB}L_{BD}}. \tag{2.7}$$

Then calculate the angle $\angle_{CAB}$ from (2.7):

$$\angle_{CAB} = (360° - \angle_{ABD} - \angle_{ACD}) - \angle_{CDB} = \angle_x - \angle_{CDB}. \tag{2.8}$$

Adding (2.8) into (2.6) results in the following formula:

$$(2L_{CD}L_{BD} - 2L_{AC}L_{AB}\cos\angle_x)\angle_{CDB} - 2L_{AC}L_{AB}\angle_x \sin\angle_{CDB} \\ = L_{BD}^2 + L_{CD}^2 - L_{AC}^2 - L_{AB}^2. \tag{2.9}$$

The constants and coefficients of (2.9) can be put in a shorter form. Using the notations: $A_\angle = 2L_{CD}L_{BD} - 2L_{AC}L_{AB}\cos\angle_x$, $B_\angle = -2L_{AC}L_{AB}\angle_x$, $C_\angle = L_{BD}^2 + L_{CD}^2 - L_{AC}^2 - L_{AB}$. Equation (2.9) is described as follows: $A_\angle \cos\angle_{CDB} + B_\angle \sin\angle_{CDB} = C_\angle$. Through the formula for transforming sums $\sin(x+y) = \sin x \cos y + \cos x \sin y$, the following relationship is obtained:

$$\sin\left(\arcsin\left(\frac{A_\angle}{\sqrt{A_\angle^2 + B_\angle^2}}\right) + \angle_{CDB}\right) = \frac{C_\angle}{\sqrt{A_\angle^2 + B_\angle^2}}. \tag{2.10}$$

From (2.10) $\angle_{CDB}$ is derived:

$$\angle_{CDB} = \arcsin\left(\frac{C_\angle}{\sqrt{A_\angle^2 + B_\angle^2}}\right) - \arcsin\left(\frac{A_\angle}{\sqrt{A_\angle^2 + B_\angle^2}}\right). \tag{2.11}$$

After calculating all of the angles in the rectangle, the tilting angle of the wheel can be determined. The sum of the following angles based on Fig. 2.3c can be written as

$$\angle_{\bar{A}CA} + \angle_{ACD} + \angle_{DC\bar{C}} = 180°, \tag{2.12}$$

where $\angle_{\bar{A}CA} = \arctan((A_y - a_y - C_y)/(A_z - C_z))$. Through the parallelism of $D\bar{D}$, $B\bar{B}$ and $C\bar{C}$, angle $\angle_{DC\bar{C}}$ can be described as

$$\angle_{CDB} = \angle_{DC\bar{C}} + (\gamma + \gamma_0), \tag{2.13}$$

where γ_0 is noted as the angle between L_{BD} and axis z, under steady-state conditions. Substituting (2.12) and (2.13), the angle of wheel tilting is described as

$$\gamma = (\angle_{\bar{A}CA} + \angle_{ACD} + \angle_{CDB}) - 180° - \gamma_0. \tag{2.14}$$

Consequently, according to (2.14), γ is determined by $\angle_{CDB}$, $\angle_{ACD}$ and $\angle_{\bar{A}CA}$, which leads to the form

$$\gamma(\angle_{\bar{A}CA}(a_y), \angle_{CDB}(a_y, \angle_{ACD}), \angle_{ACD}) = \gamma(\angle_{ACD}, a_y). \tag{2.15}$$

The value of $\angle_{ACD}$ is an unmeasured angle in the suspension, i.e., it must be expressed through measurable parameters. In practice, the vertical motion of point T can be used for this reason. The value of this motion is determined by road roughness and suspension vertical load. The relationship between t_z and $\angle_{ACD}$ requests the computation of d_z, i.e., the vertical position of D through the orientation of triangle DTB. The position of point D is calculated as

$$d_z = L_{CD}(\sin(90^\circ - \angle_{\bar{A}CA} - \angle_{ACD}) - \sin(90^\circ - \angle_{\bar{A}CA} - \angle_{ACD,0})), \quad (2.16)$$

in which the angle of $\angle_{ACD,0}$ is considered to be the steady-state value of $\angle_{ACD}$ angle, related to the unloaded state of the vehicle. Displacement T from the rotation of the triangle can be calculated as

$$t_{z,rot} = L_{DT}(\cos\angle_{\bar{D}DT} - \cos(\angle_{\bar{D}DT} - \gamma)). \quad (2.17)$$

Total displacement of T can be calculated as the sum of (2.17) and (2.16):

$$t_z = d_z + t_{z,rot}. \quad (2.18)$$

Moreover, t_y, i.e., lateral motion of T can be known as the value of half-track change. Its value characterizes the suspension system due to its impact on tire wear. Displacement t_y is calculated as

$$t_y = d_y + t_{y,rot}, \quad (2.19)$$

where the translational and rotational parts are

$$d_y = L_{CD}(\cos(90^\circ - \angle_{\bar{A}CA} - \angle_{ACD}) - \cos(90^\circ - \angle_{\bar{A}CA} - \angle_{ACD,0})), \quad (2.20)$$

$$t_{y,rot} = L_{DT}(\sin\angle_{\bar{D}DT} - \sin(\angle_{\bar{D}DT} - \gamma)). \quad (2.21)$$

Equation (2.18) shows that t_z greatly depends on both γ and $\angle_{ACD}$. Likewise, (2.19) shows that t_y depends on both γ and $\angle_{ACD}$. Using (2.15), γ depends on both $\angle_{ACD}$ and a_y, hence t_y and t_z are influenced by the same signals. Therefore, through a_y and $\angle_{ACD}$ the values of γ, t_y and t_z are computed.

2.2.3 *Iterative Solution on the Motion of Double-Wishbone Suspension*

The following equations are used to calculate the coordinate of the double-wishbone suspension points (Fig. 2.3c):

$$D_y^2 + D_z^2 = L_{DC}^2, \tag{2.22a}$$
$$(D_y - B_y)^2 + (D_z - B_z)^2 = L_{BD}^2, \tag{2.22b}$$
$$(A_y - B_y)^2 + (A_z - B_z)^2 = L_{AB}^2, \tag{2.22c}$$
$$(D_y - T_y)^2 + (D_z - T_z)^2 = L_{TD}^2, \tag{2.22d}$$
$$(B_y - T_y)^2 + (B_z - T_z)^2 = L_{TB}^2, \tag{2.22e}$$

where $(A_y, B_y, C_y, D_y, T_y, A_z, B_z, C_z, D_z, T_z)$ are the suspension point coordinates in regards to axle y, z, and L_{AB}, L_{DC} are suspension arm lengths, and L_{BD}, L_{TB}, L_{TD} denote distance values between distinct suspension points. The position of point A can be separated to $A_y = \bar{A}_y + a_y$ and $A_z = \bar{A}_z + a_z$. In addition, the coordinate values of other points are also divided as the sum of $(\bar{A}_y, \bar{B}_y, \bar{C}_y, \bar{D}_y, \bar{T}_y, \bar{A}_z, \bar{B}_z, \bar{C}_z, \bar{D}_z, \bar{T}_z)$ and $(a_y, b_y, c_y, d_y, t_y, a_z, b_z, c_z, d_z, t_z)$. The first part of suspension coordinates stands for their nominal values, which have constant values due to the steady-state vehicle position. The displacement of these points is represented by the second part. The constraints $a_z = 0$, $c_y = 0$ and $c_z = 0$ are defined in relation to the suspension motion. The camber angle of the suspension can be computed as

$$\gamma = \arccos \frac{B_z - D_z}{L_{BD}}. \tag{2.23}$$

Through the transformation of (2.22), it is possible to describe the connection between control intervention a_y and the motion of the suspension points, which are compressed into vector $\eta = \begin{bmatrix} b_y \ b_z \ d_y \ d_z \ t_y \end{bmatrix}^T$. Note that in the equation T_z represents the disturbance. As an example, to use point coordinate separation, the two components of (2.22), namely (2.22c) and (2.22d), are reordered:

$$[b_y + 2(\bar{B}_y - \bar{A}_y)]b_y + [b_z + 2(\bar{B}_z - \bar{A}_z)]b_z = [-a_y - 2(\bar{A}_y - \bar{B}_y)]a_y, \tag{2.24a}$$
$$\begin{aligned}[d_y + 2(\bar{D}_y - \bar{T}_y)]d_y + [d_z + 2(\bar{D}_z - \bar{T}_z)]d_z + [t_y + 2(\bar{T}_y - \bar{D}_y)]t_y \\ = [-t_z^2 - 2t_z(2\bar{T}_z - 2\bar{D}_y)].\end{aligned} \tag{2.24b}$$

As a consequence, (2.22) is organized in the following formula:

$$A_\eta(\eta)\ \eta = K(t_z) + B_\eta(a_y)\ a_y. \tag{2.25}$$

Furthermore, (2.23) is also arranged into the following formula:

$$C_\eta \eta = D_\eta, \tag{2.26}$$

where $C_\eta = \begin{bmatrix} 0 \ 1 \ 0 \ -1 \ 0 \end{bmatrix}^T$ and $D_\eta = [L_{BD} \cos(\gamma) + D_z - B_z]$. Lastly, vector η is derived from (2.25) in the following form:

$$\eta = A_\eta(\eta)^{-1}[K(t_z) + B_\eta(a_y)a_y]. \tag{2.27}$$

Replacing (2.27) in (2.26):

$$C_\eta A_\eta(\eta)^{-1}[K(t_z) + B_\eta(a_y)a_y] = D_\eta. \tag{2.28}$$

Based on (2.28), the a_y control input can be derived as

$$a_y = (C_\eta A_\eta(\eta)^{-1} B_\eta(a_y))^{-1}[D_\eta - C_\eta A_\eta(\eta)^{-1} K(t_z)]. \tag{2.29}$$

Equation (2.29) expresses the connection of a_y and γ, which has a parameter-dependent form, depending on the values of η, t_z and a_y. In this relationship, variables A_η, D_η and $K(t_z)$ vary as a function of η and t_z, while η is unknown. Furthermore, B_η is dependent on a_y. As a consequence, the value of a_y can be computed only through an iterative way.

The solution of the iteration requests to have information about t_z road excitation value, and, moreover, initial values of η_0, a_{y0} must be selected. Initially, the new value for η is computed based on (2.27). Next, input a_y is calculated based on (2.29), then finally the mean square of the error is given as

$$\vartheta = \sum((\eta - \eta_0)^2 + (a_y - a_{y0})^2). \tag{2.30}$$

The termination condition for stopping the iteration is defined as $\vartheta < \varepsilon$, where ε is a preselected small scalar value. Otherwise the re-iteration is accomplished by choosing $a_{y0} = a_y$ and $\eta_0 = \eta$. Some procedures for numerical solutions can be found in Kelley (1995) and Coleman and Li (1994).

2.2.4 Model Formulation for McPherson Suspensions

The illustration of the McPherson suspension is found in Fig. 2.3b. In the case of McPherson suspension constructions, the intervention of the system is the y lateral motion of point C, i.e., c_y. Here, C_y is the lateral coordinate of point C and $C_{y,0}$ is the stead-state position coordinate, which results in $C_y = C_{y,0} + c_y$. The connection point of the road and tire is marked with point T.

The upper connection point of the damper is noted with A, which in the case of a real construction is a Silent-block unit. This unit is treated as a joint, like C and D. The other strut end is connected to the hub directly, without a joint. Point B can be considered as a virtual point, which is depicted by the intersection of the wheel hub and suspension strut alignment. T is the road–wheel contact point, which has two causes for its movement. The first reason is due to road irregularities. Since the coordinate system is attached to the chassis, chassis roll motion has a similar impact. w_y and w_z note these movements. Secondly, the operation of the control must be considered. In accordance with c_y intervention, tire position is also varied, which results in vertical and lateral movements of T, i.e., t_y and t_z. The sums of the movements are $T_z = w_z + t_z$, $T_y = w_y + t_y$.

The aim of the model formulation is to specify the connection between the input c_y, the output of the wheel camber γ and the displacement of point T. In the calculation of the suspension point positions, rectangle $ABDC$ is examined. All sides of this rectangle are considered to be known, i.e., L_{AB} results in the measured suspension deflection, L_{CD}, L_{BD} are constants and L_{AC} is computed by knowing the positions of A and C. The control input c_y is known, the point A position is fixed and the vertical position of C is also constant. This means that each side of the rectangle is calculated, and the positions B and D must be calculated.

Since the wheel hub and the strut in point B are connected, angle $\angle_{ABD}$ can be handled as constant. Knowing the value of $\angle_{ABD}$ and using the law of cosine, L_{AD} is calculated:

$$L_{AD}^2 = L_{AB}^2 + L_{BD}^2 - 2L_{AB}L_{BD}\cos\angle_{ABD}. \tag{2.31}$$

Taking into account triangle ACD, the change in $\angle_A CD$ is calculated according to the law of the cosine:

$$\angle_{ACD} = \frac{\arccos(L_{AC}^2 + L_{CD}^2 - L_{AD}^2)}{2L_{AC}L_{CD}}. \tag{2.32}$$

Through the computed angles and the side lengths, the positions of B, D points can be calculated. When c_y is actuated and the suspension is deflected, the orientation of suspension arm CD around C is modified. The modification in the position of point D can be calculated as follows:

$$\overrightarrow{AD} = \left(\begin{bmatrix} 1 & 0 \\ 0 & 1 \end{bmatrix} + \frac{L_{CD}}{L_{AC}} \cdot \begin{bmatrix} \cos(\pi - \angle_{ACD}) & -\sin(\pi - \angle_{ACD}) \\ \sin(\pi - \angle_{ACD}) & \cos(\pi - \angle_{ACD}) \end{bmatrix}\right) \overrightarrow{AC}. \tag{2.33}$$

By adding the vector $\overrightarrow{AD}$ to A point, the vertical and lateral positions of D can be calculated. The coordinates of point B are calculated with the same method, such as

$$\overrightarrow{AB} = \left(\frac{L_{AB}}{L_{AD}} \cdot \begin{bmatrix} \cos(\angle_{DAB}) & -\sin(\angle_{DAB}) \\ \sin(\angle_{DAB}) & \cos(\angle_{DAB}) \end{bmatrix}\right) \overrightarrow{AD}, \tag{2.34}$$

in which the expression $\angle_{DAB}$ can be calculated through the formula of the law of cosines:

$$\angle_{DAB} = \frac{\arccos(L_{AD}^2 + L_{AB}^2 - L_{BD}^2)}{2L_{AD}L_{AB}}. \tag{2.35}$$

Through the addition of vector $\overrightarrow{AB}$ to point A, the vertical and lateral positions of B are computed.

Finally, the camber angle must be determined. It is calculated with coordinates B and D:

$$\gamma = \arctan \frac{B_y - D_y}{B_z - D_z}. \tag{2.36}$$

By using a rotation matrix, the tire–road contact point T can be calculated:

$$\overrightarrow{DT} = \left(\begin{bmatrix} \cos\gamma & -\sin\gamma \\ \sin\gamma & \cos\gamma \end{bmatrix} \right) \overrightarrow{DT}_0, \tag{2.37}$$

where $\overrightarrow{DT}_0$ is the vector determined in the steady-state position of the vehicle. By adding $\overrightarrow{DT}$ to the coordinates of D, the coordinates of T are calculated, which can have importance due to tire wear (Gough and Shearer 1956). Knowing the new coordinate of T, the change of half-track can be calculated as $\varDelta B = T_y - T_{y,0}$, in which relationship $T_{y,0}$ represents the lateral position of the contact point under steady-state conditions.

2.2.5 *Interactions Between Different Motions in Variable-Geometry Suspension*

The interactions between different motions have been described for a double-wishbone suspension below. In spite of the given construction, the method of the analysis can be extended to the examination of the McPherson type of suspension.

By changing the tilting angle, the entire suspension geometry is also modified, which affects the rotation of the wheel, i.e., γ induces rotation of the wheel, whose orientation is varied around an axis defined by D (attachment of the lower arm) and K (steering track-rod end). In this manner, lateral and vertical positions of point K play a role in the rotation of the wheel. The angle between BK axis and the road plane determines the relationship between γ and δ_c. Besides, a steering angle addition through γ during the suspension operation is generated. As a consequence, the proper suspension geometry can improve the lateral force of the tire not only with the camber angle but also with the steering angle (Németh and Gáspár 2013, 2012a).

The wheel position can be calculated as follows. The vertical and lateral motions of B, D points, i.e., b_z, d_z, b_y, d_y, are defined by a_y control input and t_z disturbance. Furthermore, the rotation of the wheel is influenced by the position of K. The orientation of plane BDK in the suspension is characterized by vector $\overrightarrow{N}$. Thus, the values of δ_c and γ can be calculated through $\overrightarrow{N}$, such as

$$\delta_c = \arctan\left(\frac{N_x}{N_y}\right), \tag{2.38a}$$

$$\gamma = \arctan\left(\frac{N_z}{\sqrt{N_x^2 + N_y^2}}\right), \tag{2.38b}$$

in which expressions N_x, N_y and N_z represent the components of $\overrightarrow{N}$. It can be calculated as

$$\overrightarrow{N} = \overrightarrow{DB} \times \overrightarrow{DK}, \tag{2.39}$$

where $\overrightarrow{DB}$, $\overrightarrow{DK}$ are vectors between track-rod end and suspension points, see details in Németh and Gáspár (2011c), i.e., calculation of the positions of B, D require measurement on suspension compression and a_y.

When calculating point K, the following statements must be taken into account in the directions x, y and z:

- Steering axis in a double-wishbone type of suspension is defined by axis BD.
- It is considered a rigid solid wheel hub, which induces that distances between the points of K, D and B are constant.
- Steering track-rod is considered to be rigid, i.e., its length has constant value.

These assumptions provide that lateral and vertical positions of K can be computed, if the steering rack position and the positions of B and D are known. Since the steering rack only in the lateral direction can be moved, the positions at the end of steering rack S_x, S_z have constant values, and S_y by the steering input is determined. Thus, the measurement of S_y is requested.

The coordinate of K can be computed in several ways. An analytic solution can be led by the model to consider K as a point, which is in the intersection of 3 balls. Since the segments of SK, DK and BK are considered to be constant, the positions of the points B, D and S result in the following coordinate-geometry-based relations:

$$BK^2 = (K_x - B_x)^2 + (K_y - B_y)^2 + (K_z - B_z)^2, \tag{2.40a}$$

$$DK^2 = (K_x - D_x)^2 + (K_y - D_y)^2 + (K_z - D_z)^2, \tag{2.40b}$$

$$SK^2 = (K_x - S_x)^2 + (K_y - S_y)^2 + (K_z - S_z)^2. \tag{2.40c}$$

There are three unknown variables (K_x, K_y, K_z) in the equations, which are the positions of point K. Though (2.40) results in an analytic solution, its computation can lead to some numerical difficulties.

Another solution for the computation of suspension point positions is to model the construction via Simscape Multibody™, as depicted in Fig. 2.4. Arms and bodies of the suspension are connected by joints to the chassis of the vehicle. The joints A_1 and A_2 are actuated laterally, altering the position and direction of the wheel. While the point coordinates during the simulations are measured, values of δ_c and γ are computed through (2.38).

Fig. 2.4 Modeling suspension system through SimMechanics

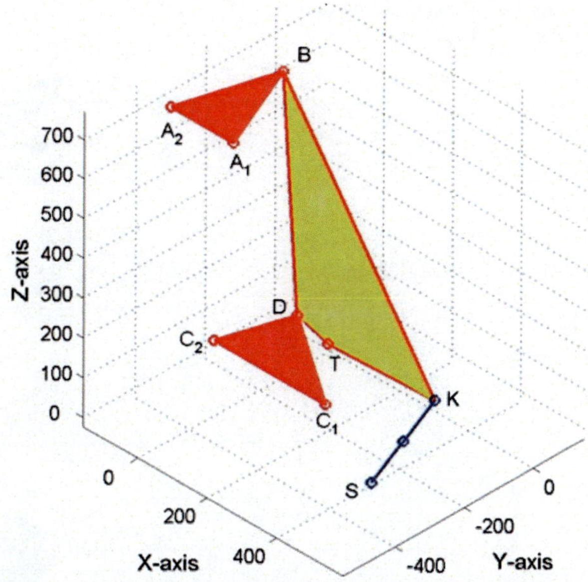

Figure 2.5 shows the connection between each signal, i.e., δ_c, γ angles and ΔB for different K_z values. It illustrates that the variation of K_z can have a large effect on δ_c and a slight influence on γ, due to the orientation of the axis of KB. Since signals δ_c, γ are usually conflicting, an appropriate solution must be found for the setting of parameter K_z. In the detailed design, K_z has a big effect on δ_c, and with increased K_z high lateral tire force can be achieved; see Sect. 2.1. In addition, K_z affects t_y, such as half-track change ΔB. As a consequence, half-track change, tilting and steering angle are functions of a_y, i.e., $\Delta B = \Delta B(a_y)$, $\gamma = \gamma(a_y)$ and $\delta_c = \delta_c(a_y)$, whose relationships through a fitting process on the results of various simulations can be formed.

2.3 Examination on the Motion Characteristics of Variable-Geometry Suspension

Next, the impact of a_y on the motion characteristics of a double-wishbone type variable-geometry suspension is examined. The connection between a_y and γ, depending on t_z, through a function can be described; see an exemplary illustration in Fig. 2.6a. This examination illustrates the possibility of fitting linear curves:

$$\gamma = \kappa + \xi_1 t_z + \varepsilon_1 a_y. \tag{2.41}$$

Fig. 2.5 Effect of K_z on the signals ΔB, γ and δ_c

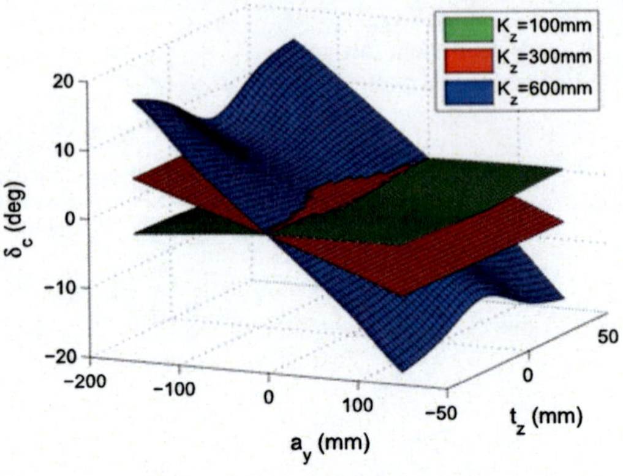

(a) δ_c change

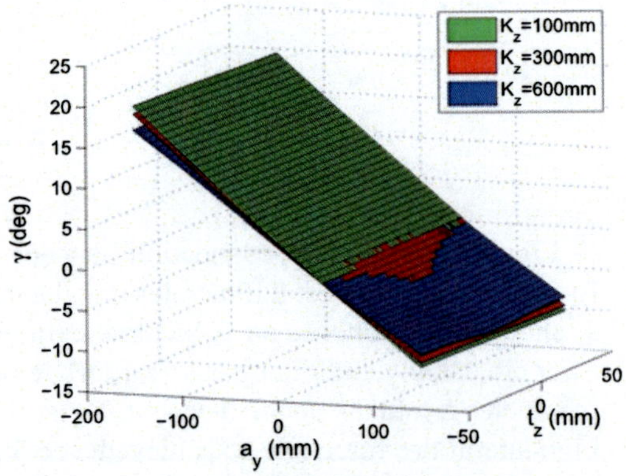

(b) γ change

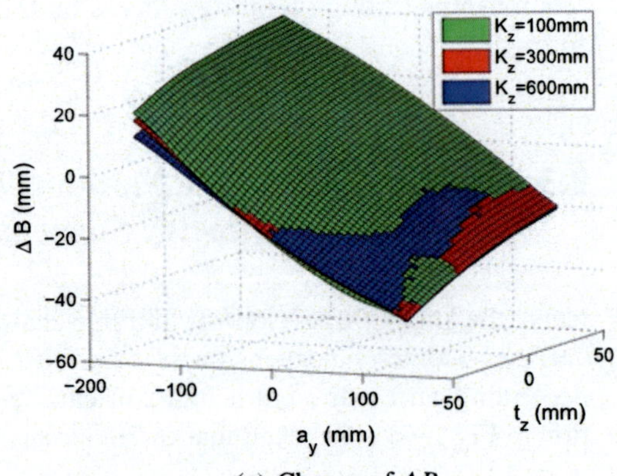

(c) Change of ΔB

Fig. 2.6 Variable-geometry vehicle suspension motion characteristics (©2012 IEEE. Reprinted, with permission, from Németh and Gáspár 2011b)

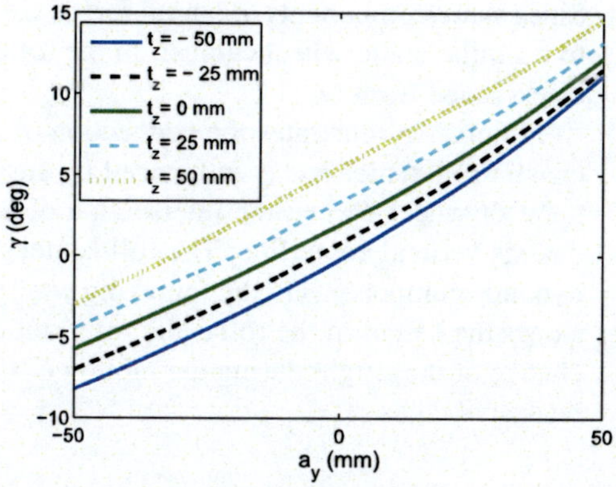

(a) a_y - γ characteristics

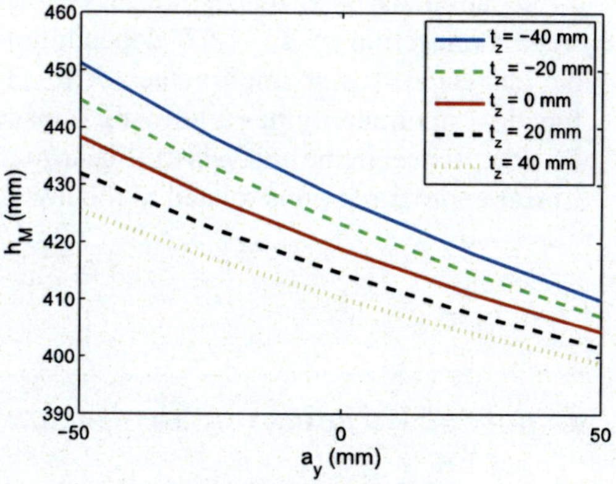

(b) Roll center height

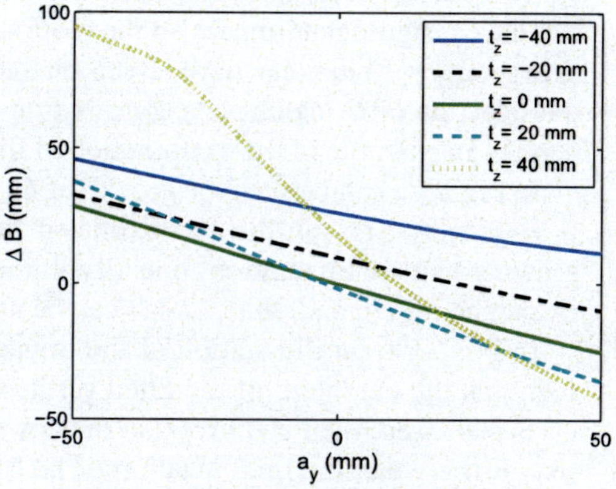

(c) Change of half track

Since static components of lateral forces are approximately equal on each wheel due to a similar static wheel camber, in the following calculations κ constant value can be excluded from (2.41).

Figure 2.6b illustrates the connection of $a_y - h_M$, depending on t_z. The value of chassis roll center h_M is influenced by suspension points A, B, C, D and also the point of wheel T, i.e., the intersection of the lines of the suspension arms and the chassis vertical centerline. The roll center height can be partitioned into static and dynamic components in the following way: $h_M = h_{M,st} + \Delta h_M$. The element $h_{M,st}$ shows the height of the roll center of a stationary vehicle, while Δh_M illustrates the change of the height during the motion of the vehicle. The dynamic component can be described as

$$\Delta h_M = \xi_2 t_z + \varepsilon_2 a_y. \tag{2.42}$$

Variation ΔB is also a significant character of the suspension motion due to tire wear. Connection of $a_y - \Delta B$ depending on t_z is illustrated in Fig. 2.6c. Though the characteristics at small values of (a_y, t_z) can be approximated through linear function, nonlinearity has a growing impact on half-track change at high values of (a_y, t_z). Hence, in the analysis the linear/nonlinear components of ΔB are separated. Linear approximation is related to the low range, such as

$$\Delta B = \xi_3 t_z + \varepsilon_3 a_y. \tag{2.43}$$

Impact of Variations in Suspension Construction Parameters

In this subsection, the impact of variations in suspension parameters is studied, especially the effect of C_z selection on the previous characteristics. Since the construction can have a significant impact on the motion characteristics, the selection of suspension geometry values can have effects on the control intervention. Thus, for example, the selection of C_z can have a contribution to the control design process. Figure 2.7 depicts an example of the examination of the characteristics for different C_z values.

In Fig. 2.7a can be seen the growth of C_z results in a growing gradient of function $\gamma = f(a_y, t_z)$, i.e., in the approximated function (2.41) the parameters ξ, ε have growing values. In addition, linearity in the characteristics of $\gamma = f(a_y, t_z)$ for all C_z can be found, and, thus, (2.41) can be applied in the given range of C_z.

In Fig. 2.7b, an illustration of the impact of C_z on ΔB can be found. It can be seen that the curvature of the fitted surface extends for higher C_z values. Hence, if C_z is increased, then ΔB grows at the down motion of the suspension. To avoid the risk of tire wear, the minimization of $|\Delta B|$ must be guaranteed. One possible way to minimize ΔB is to limit a_y. Another way to minimize the half-track change is to choose the right position for the lower arm.

Figure 2.7c illustrates that C_z greatly affects h_M. Its reason is that C_z has a high impact on the value of h_M; see the work of Mitschke and Wallentowitz (2004)

Fig. 2.7 Illustration of the impact of C_z selection

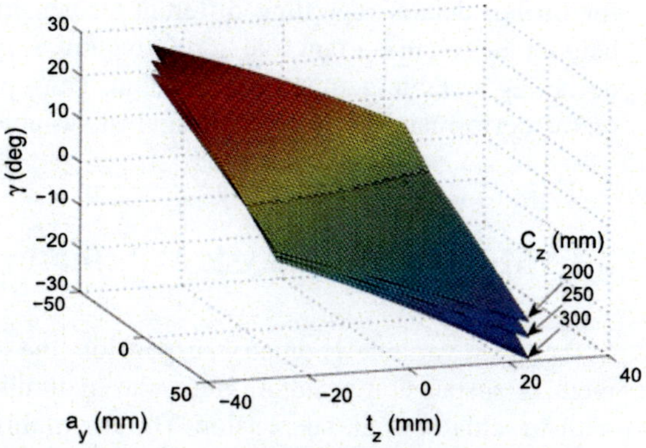

(a) Effect on γ

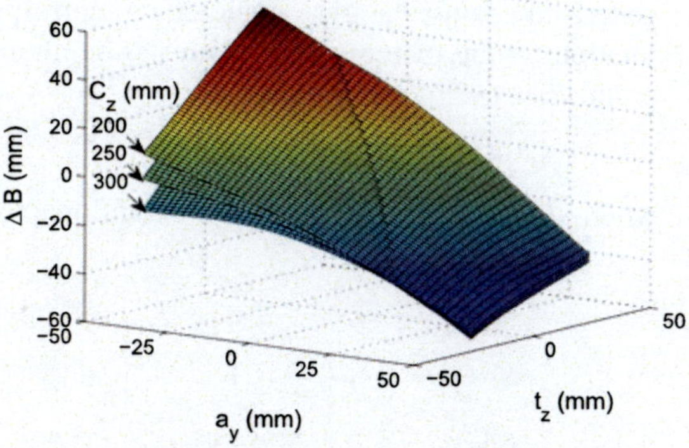

(b) Effect on half-track change

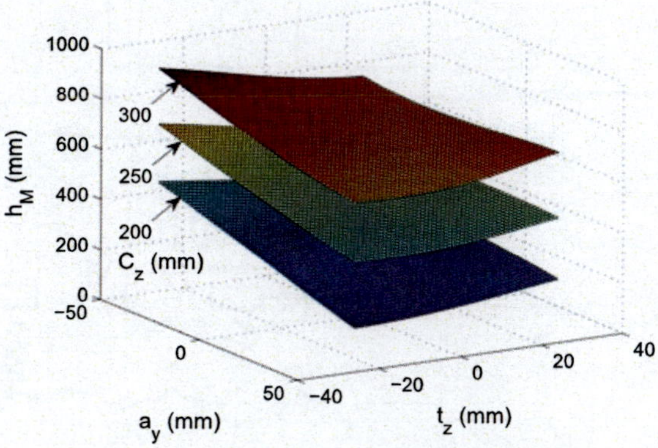

(c) Effect on height of the roll center

for further details regarding different suspension constructions. Since roll center height plays an important role in roll dynamics, h_M variation can be important when designing variable-geometry suspension. Thus, proper selection of C_z has an impact on the performance requirements against variable-geometry suspension control.

2.4 Mechanical Analysis of Actuator Intervention

In the given variable-geometry suspension, the control input is the lateral displacement of suspension point A. Here, the hydraulic structure is considered to be the built-in actuator of the suspension. The aim of this analysis is to compute the required actuator force in order to realize a_y.

For generating γ, different resistances must be compensated through the hydraulic actuator of the suspension. To change the tilting of rotated wheels, energy must be generated against the gyroscopic effect, corresponding to the steering system. The rotation torque of the wheel about its longitudinal axis is given by the following premise:

$$M_{gy} = 2J_w v / r_w \dot{\gamma}, \tag{2.44}$$

where J_w is the wheel inertia on the rotation axis and r_w is the radius of the wheel.

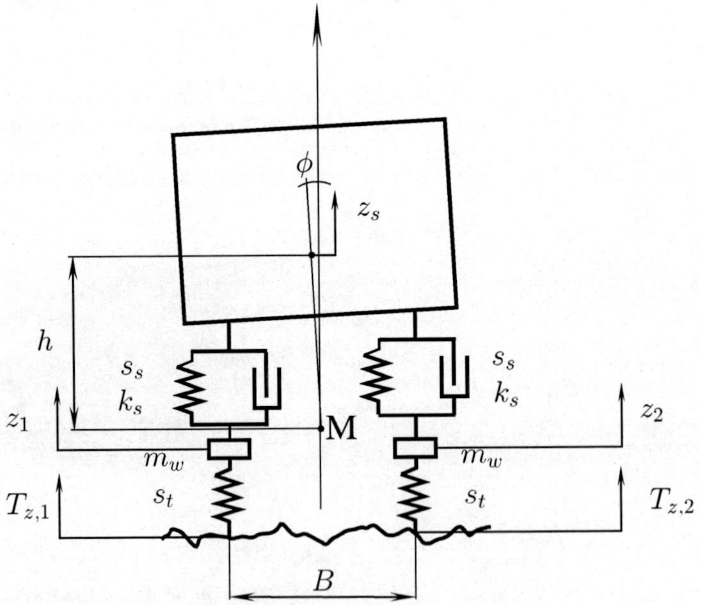

Fig. 2.8 Illustration of vehicle roll dynamical model

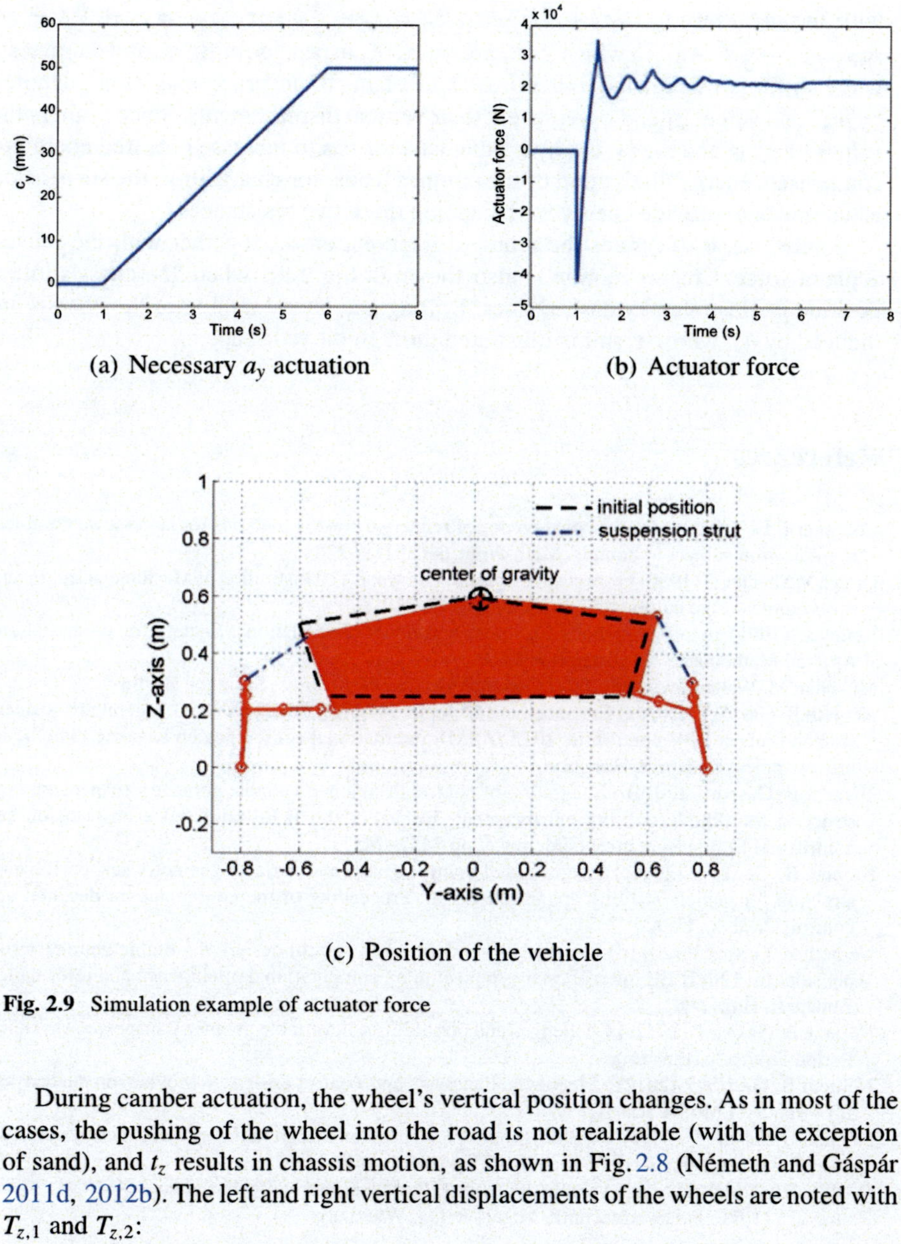

Fig. 2.9 Simulation example of actuator force

During camber actuation, the wheel's vertical position changes. As in most of the cases, the pushing of the wheel into the road is not realizable (with the exception of sand), and t_z results in chassis motion, as shown in Fig. 2.8 (Németh and Gáspár 2011d, 2012b). The left and right vertical displacements of the wheels are noted with $T_{z,1}$ and $T_{z,2}$:

$$(I_{xx} + mh^2)\ddot{\phi} = mgh\phi - s_s z_p - k_s z_{pd} + s_s z_m + k_s z_{md}, \tag{2.45a}$$

$$m\ddot{z}_s = s_s(z_1 - z_s) + s_s(z_2 - z_s) + k_s(\dot{z}_1 - \dot{z}_s) + k_s(\dot{z}_2 - \dot{z}_s), \tag{2.45b}$$

$$m_w\ddot{z}_1 = s_t(T_{z,1} - z_1) + s_s z_p + k_s z_{pd}, \tag{2.45c}$$

$$m_w\ddot{z}_2 = -s_t(T_{z,2} - z_2) + s_s z_m + k_s z_{md}, \tag{2.45d}$$

with the notations $z_p = z_s + \frac{B}{2}\phi - z_1$, $z_m = z_s - \frac{B}{2}\phi - z_1$, $z_{pd} = \dot{z}_s + \frac{B}{2}\dot{\phi} - \dot{z}_1$, $z_{md} = \dot{z}_s - \frac{B}{2}\dot{\phi} - \dot{z}_1$, in which s_s, k_s are ratios of suspension stiffness and dampness, while stiffness s_t is related to the tire. The weights of unsprung masses are denoted by m_w, and values z_1 and z_2 represent their vertical displacements. Since t_z alters the values of z_s, ϕ and z_1, z_2, the hydraulic actuator has to increase potential energy to compensate energy dissipation of the damper. Thus, for changing γ, the suspension actuator has to provide energy for balancing these two resistances.

Figure 2.9a, b illustrates the required intervention a_y, together with the related actuator force. Chassis motion is also shown in Fig. 2.9c, where the chassis initial position is illustrated by dashed lines. Rolling and lateral motion of the chassis are induced by a_y, whose result is illustrated through the red shape.

References

Coleman T, Li Y (1994) On the convergence of reflective newton methods for large-scale nonlinear minimization subject to bounds. Math Progr 64(2):189–224

Gough V, Shearer G (1956) Front suspension and tyre wear. The Institution of Mechanical Engineers, Proceedings of the automobile division, pp 171–216

Kelley C (1995) Iterative methods for linear and nonlinear equations. Society for Industrial and Applied Mathematics, Philadelphia, USA

Mitschke M, Wallentowitz H (2004) Dynamik der Kraftfahrzeuge. Springer, Berlin

Németh B, Gáspár P (2011a) Enhancement of vehicle stability based on variable geometry suspension and robust LPV control. In: IEEE/ASME international conference on advanced intelligent mechatronics, Budapest, Hungary

Németh B, Gáspár P (2011b) Integration of control design and variable geometry suspension construction for vehicle stability enhancement. In: 2011 50th IEEE conference on decision and control and European control conference, pp 7452–7457

Németh B, Gáspár P (2011c) Integration of control design and variable geometry suspension construction for vehicle stability enhancement. In: Proceeding of the conference on decision and control, Orlando, USA

Németh B, Gáspár P (2011d) Uncertainty modeling and control design of variable geometry suspension. In: 12th IEEE international symposium on computational intelligence and informatics, Budapest, Hungary

Németh B, Gáspár P (2012a) Challenges and possibilities in variable geometry suspension systems. Period Polytech Trans Eng

Németh B, Gáspár P (2012b) Mechanical analysis and control design of McPherson suspension. Int J Veh Syst Model Test 7(2):173–193

Németh B, Gáspár P (2013) Variable-geometry suspension design in driver assistance systems. In: 2013 European control conference (ECC), pp 1481–1486

Pacejka HB (2004) Tyre and vehicle dynamics. Elsevier Butterworth-Heinemann, Oxford

Reimpell J (1976) Fahrwerktechnik. Vogel-Verlag, Würzburg

Chapter 3
LPV-Based Control of Variable-Geometry Suspension

This chapter deals with the design of variable-geometry suspension, i.e., suspension structure along with the control design and their interactions are examined, while a simultaneous design method is introduced.

3.1 Performances of Variable-Geometry Suspension Systems

Roll and lateral dynamics of the vehicle can be effectively influenced by variable-geometry vehicle suspensions, and by further active control interventions, e.g., differential braking. Nevertheless, the performances of these dynamics depend not only on the control system but also on the construction. Note that this controlled system operates simultaneously with steering and differential braking systems. The performance requirements, which determine the design of the control and the construction, are as follows (Németh and Gáspár 2013, 2012a):

- Path following capability of the driver can be enhanced through variable-geometry vehicle suspensions. This functionality requests the tracking of a reference signal on yaw rate, which leads to the criterion

$$|\dot{\psi}_{ref} - \dot{\psi}| \rightarrow min. \tag{3.1}$$

- Improvement of the roll dynamics can be achieved through the reduction of the roll angle of the vehicle chassis. In the case of variable-geometry suspension, the vertical position of the roll center has an impact on the roll dynamics; see Gáspár and Bokor (2005). The motion of the vehicle, i.e., lateral acceleration from a planar viewpoint is formed as

B. Németh and P. Gáspár, *Control of Variable-Geometry Vehicle Suspensions*, Advances in Industrial Control, https://doi.org/10.1007/978-3-031-30537-5_3

$$\bar{a} = \bar{u}_R \left(\ddot{R}(t) - \frac{v^2(t)}{R(t)} \right) + \bar{u}_\perp \left[\frac{2}{R(t)} \dot{R}(t)v(t) + R(t)\frac{d}{dt}\left(\frac{v(t)}{R(t)}\right) \right], \quad (3.2)$$

where $\bar{u}_R$ is radial and $\bar{u}_\perp$ is angular unit vector, with regard to circular motion. $v(t)$ denotes the longitudinal speed of the vehicle, and circular motion radius is $R(t)$. The centripetal acceleration influences roll dynamics, accordingly, $\bar{u}_R$ is contemplated as

$$(I_{xx} + m\Delta h^2)\ddot{\phi} = mg\Delta h\phi + m\left(\ddot{R}(t) - \frac{v^2(t)}{R(t)}\right)\Delta h - B_i \sum_{i=1}^{4} F_{susp,i}, \quad (3.3)$$

where Δh expresses the distance between roll center and chassis CoG ($\Delta h = h_{CG} - h_M$). Moreover, ϕ angle is related to the roll of the chassis, the half-track is B_i and chassis inertia is denoted by I_{xx}. In (3.4), the vertical forces in the suspension system are denoted by $F_{susp,i}$. Thus, through the modification of Δh, the roll angle can be influenced. Note that roll dynamics is influenced, because of the relation $\ddot{R}(t) - \frac{v^2(t)}{R(t)}$. Assuming that $\ddot{R}(t)$ has a low value, together with the relation on the rest of lateral acceleration, $\frac{v^2(t)}{R(t)} = v(t)(\dot{\beta}(t) + \dot{\psi}(t))$, roll dynamics is given as follows:

$$(I_{xx} + m\Delta h^2)\ddot{\phi} = mg\Delta h\phi + mv\Delta h(\dot{\beta} + \dot{\psi}) - B_i \sum_{i=1}^{4} F_{susp,i}. \quad (3.4)$$

The reduction of roll angle can be achieved through the minimization of roll center height h_M. It can be separated into static and dynamic terms, such as $h_M = h_{M,st} + \Delta h_M$. It leads to the criterion that the distance between centers of gravity and roll has to be reduced (3.5a), and the dynamic motion of the roll height should be minimized:

$$|\Delta h_M| = |\xi_2 t_z + \varepsilon_2 a_y| \rightarrow min, \quad (3.5a)$$

$$|h_{CG} - h_{M,st}| \rightarrow min. \quad (3.5b)$$

In terms of performance of (3.5a), it can be seen that the height of the roll center in the steady state is defined by the structure of the suspension. Equation (3.5b) describes that the vertical movement of the roll is given by t_z and a_y. This suggests that the reduction in height of the roll center is influenced by suspension construction and control simultaneously.

- Half-track change of the wheels is another important economic factor. The lateral movement of the contact point T is relevant in the aspect of tire wear (Gough and Shearer 1956), when the suspension moves vertically while the vehicle moves forward. With adequate variable-geometry control, unwanted lateral tire motion can be eliminated. Through (2.43), the following performance criterion is formulated:

$$|\Delta B| = |\xi_3 t_z + \varepsilon_3 a_y| \rightarrow min. \tag{3.6}$$

Nevertheless, the linear approximation of ΔB can lead to imperfections in the nonlinear region of the suspension motion. Therefore, ΔB on an increased region through integration can be summed up as

$$\int\int |\Delta B| da_y dt_z \rightarrow min. \tag{3.7}$$

- Large control inputs must be prevented during control tasks due to the fact that a_y has design constraints. Accordingly, the formulation of the performance criterion can lead to the reduction of the intervention on the control motion:

$$|a_y| \rightarrow min. \tag{3.8}$$

The formulated performance criteria are separated into the following group parts: one depends on the control system and the other on the structure of the suspension. The performances in the first group, i.e., (3.1), (3.5a), (3.6) and (3.8) are formed through a_y and the vertical motion t_y, in a linear expression. These criteria in a $\mathscr{Z}_1$ vector are formed, which are used in the control design process:

$$\mathscr{Z}_1 = \begin{bmatrix} \dot{\psi}_{ref} - \dot{\psi} \\ \xi_2 t_z + \varepsilon_2 a_y \\ \xi_3 t_z + \varepsilon_3 a_y \\ a_y \end{bmatrix}. \tag{3.9}$$

The (3.5b) and (3.7) performances cannot be described in a linear variation and are only used as a part of the design process of the suspension construction. Thus, in $\mathscr{Z}_2$ vector the following signals are involved:

$$\mathscr{Z}_2 = \begin{bmatrix} h_{CG} - h_{M,st} \\ \int\int |\Delta B| da_y dt_z \end{bmatrix}. \tag{3.10}$$

3.2 Optimization of Vehicle Suspension Constructions

The design of the optimal construction is motivated by the effect of the structure of the suspension on the operation of the system, i.e., the performance criteria in Sect. 3.1 are influenced by control and construction simultaneously. A method for the joint design process is given in this section (Németh and Gáspár 2013). This interconnection is the cause that the structure of the system affects the characteristics $\gamma(a_y, t_z)$, ΔB and Δh_M, which are related to the elements of control design (3.9), too. Performance of (3.10) also influences vehicle dynamics. Variables of the suspension system $\Phi_1 \ldots \Phi_n$ must be considered in the construction design. Note that $\Phi_i, i \in [1, n]$ value has physical limitations, due to the structure of the vehicle chassis (e.g., wheel size and hub): $\bar{\Phi}_i = [\Phi_{i,min}\ \Phi_{i,max}], i \in [1, n]$. Simultaneous design is accomplished in the following phases.

First, $\mathscr{T}_{z,j}(\Phi_i)$ operator calculates $\mathscr{Z}_1$ norm values. The calculations show the achieved $\mathscr{Z}_1$ performance level in case of various suspension constructions:

$$J_{z,j}(\Phi_1 \dots \Phi_n) = \|\mathscr{T}_{z,j}(\Phi_1 \dots \Phi_n)\|, \quad j \in [1, 4]. \tag{3.11}$$

Second, the performance of $\mathscr{Z}_2$ in each construction can be calculated. Performances $\mathscr{Z}_2$ have weights for minimizing ΔB and the angle of the chassis roll. The values of the performance levels are scaled through a normalization factor, i.e., by applying their maximum values, which are multiplied with the weights $W_{p,k}$:

$$J_{z,k}(\Phi_1 \dots \Phi_n) = W_{p,k} \cdot |z_k(\Phi_1 \dots \Phi_n)|/z_{k,max}, \quad k \in \{5, 6\}. \tag{3.12}$$

The calculations show the performance level on $\mathscr{Z}_2$, for various suspension constructions.

Third, these performance values are aggregated, resulting in a total cost value for performing the performance of $\mathscr{Z}_1$ and $\mathscr{Z}_2$ in various suspension constructions, such as

$$\mathscr{J}_z(\Phi_1 \dots \Phi_n) = \sum_j J_{z,j}(\Phi_1 \dots \Phi_n) + \sum_k J_{z,k}(\Phi_1 \dots \Phi_n). \tag{3.13}$$

Finally, it is needed to find $\Phi_1 \dots \Phi_n$ construction parameters by which $\mathscr{J}(\Phi_1 \dots \Phi_n)$ is minimized. The defined optimization task is given as follows:

$$\inf_{\Phi_1 \in \bar{\Phi}_1} \inf_{\Phi_n \in \bar{\Phi}_n} \mathscr{J}_z(\Phi_1 \dots \Phi_n). \tag{3.14}$$

The solution of the optimization (3.14) using an interior point method can be achieved (Kelley 1995), which leads to an approximation of the solution through preconditioned conjugate gradients. Note that the same weighting functions are used to design the control during the optimization process.

3.3 Formulation of Weighting Functions for Control Design

The various types of performance can require to find a balance between them. In the case of variable-geometry suspension control, the formulation of weighting functions with parameter dependence is an efficient tool for this purpose. The performances of the suspension are affected by two signals, ρ_{susp} and h_{ref}. The following weighting functions stand for the importance between performances (Németh and Gáspár 2012b):

$$W_{susp,e\dot{\psi}} = \rho_{susp}/e_{\dot{\psi}_{max}}, \quad (3.15a)$$
$$W_{susp,\Delta h} = (1 - \rho_{susp})/\phi_{max}, \quad (3.15b)$$
$$W_{susp,\Delta B} = (1 - \rho_{susp})/\Delta B_{max}, \quad (3.15c)$$

where the signal ρ_{susp} is defined as a scheduling variable. Moreover, the maximum values of roll angle, half-track change and yaw rate error are denoted by ΔB_{max}, ϕ_{max} and $e_{\dot{\psi}_{max}}$. The selection of the scheduling variable is influenced by the vehicle dynamic scenario; see below.

- Under normal vehicle dynamic conditions, the purpose of a vehicle control system is to minimize ΔB. It is achieved through the selection of $\rho_{susp} = 0$, $h_{ref} = h_M$. It is considered to be the initial setting.
- In the case of a significant increase of ϕ, the vehicle control has to minimize chassis roll motion. It can be accomplished by choosing $\rho_{susp} = 0$, $h_{ref} = h_{ref,max}$.
- In the case of a critical vehicle maneuver with increased error on yaw rate, the goal of the vehicle control is to reduce $e_{\dot{\psi}}$, which has priority against the previous performance requirements. It is achieved through the setting of $\rho_{susp} = 1$ scheduling variable value and $h_{ref} = h_M$.

Nevertheless, there exists a selection on ρ_{susp}, with which a balance between the first and the second performance criteria can be achieved. For example, with the selection of $\rho_{susp} = 0$, the value of h_{ref} can be chosen within the range $h_M < h_{ref} < h_{ref,max}$

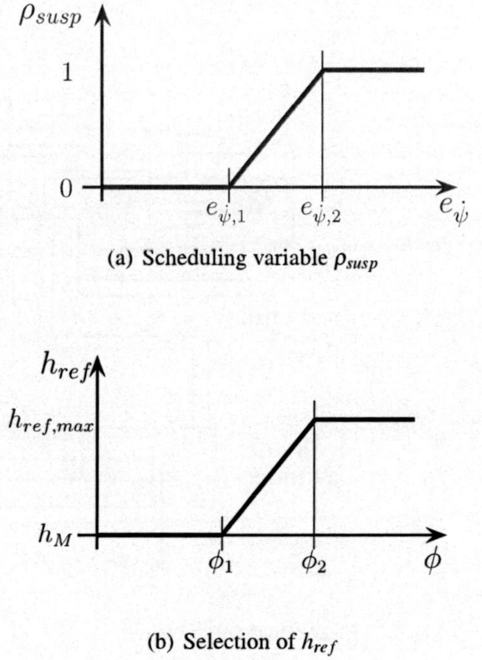

Fig. 3.1 Selection strategy on the values of h_{ref}, ρ_{susp}

to find a balance between the improvement of roll motion and half-track change. In Fig. 3.1, the selection strategy ρ_{susp} and the value of h_{ref} are illustrated, where ϕ_1, ϕ_2, $e_{\dot{\psi},1}$ and $e_{\dot{\psi},2}$ are design parameters.

3.4 Robust Control Design for Suspension Actuator

For the actuation of the variable-geometry vehicle suspension, the intervention, i.e., the lateral motion a_y must be realized. The aim of the actuator is to achieve the desired movement a_y, which indicated the setting of the actuator position. Here, an electro-hydraulic construction is considered for the suspension. Some articles consider modeling and control design for electro-hydraulic actuators, for example, Šulc and Jan (2002), Karpenko and Shapehri (2005) and Wijnheijmer et al. (2006). The system consists of an electronically controlled spool valve and a hydraulic cylinder; see below.

3.4.1 Modeling of the Hydraulic Actuator

In Fig. 3.2, the illustration of the electro-hydraulic type of suspension actuator is found. The intervention a_y is accomplished through hydraulic cylinder motion x_c, regulated by the p_L pressure variation between V_1, V_2 chambers. Pressure variation

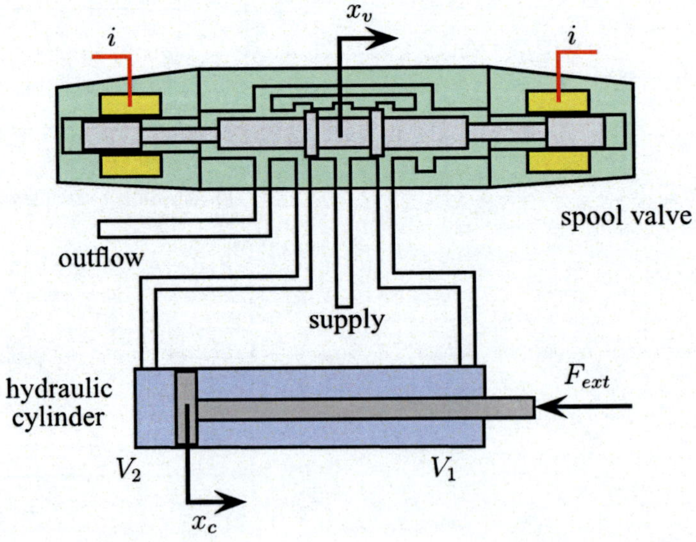

Fig. 3.2 Electro-hydraulic actuator

value and direction are affected through the movement of the spindle valve, which is regulated by the current on the armatures. Hence, the real intervention of the system is current i, and the movement x_c can be handled as an output.

In the chambers, the pressure relies upon circuit flows Q_1, Q_2. Due to the variation in the directions of the flow, from the viewpoint of control theory, the system hydraulic cylinder as a switching process can be handled. In the case of small x_c values, the formulation of the average flow is

$$Q_L = C_d A(x_v) \sqrt{\frac{1}{\rho}\left(p_s - \frac{x_v}{|x_v|} p_L\right)}, \tag{3.16}$$

which around the cylinder center position can be linearized (see Karpenko and Shapehri 2005) as follows:

$$Q_L = K_q x_v - K_c p_L, \tag{3.17}$$

where K_q is the gain coefficient, while the pressure coefficient is denoted by K_c. The differential pressure can be described using the following dynamical equation (Meritt 1967):

$$\dot{p}_L = \frac{4\beta_e}{V_t}\left(Q_L - A_p \dot{x}_c + c_{l1} \dot{x}_c - c_{l2} p_L\right), \tag{3.18}$$

where total volume under pressure is represented by V_t and effective bulk modulus is denoted by β_e. Moreover, piston area and further construction parameters are denoted by A_p, c_{l1}, c_{l2}.

According to the pressure difference and load, the cylinder position is defined by the piston motion equation:

$$m_c \ddot{x}_c + d_c \dot{x}_c = A_p p_L + F_{ext}, \tag{3.19}$$

in which d_c represents the damping coefficient and piston mass value is denoted by m_c.

The model of the spool valve can be formed as a second-order linear system that expresses the relationship among the spool movement x_v and the current i as detailed in Šulc and Jan (2002):

$$\frac{1}{\omega_v^2} \ddot{x}_v + \frac{2d_v}{\omega_v} \dot{x}_v + x_v = k_v i, \tag{3.20}$$

in which ω_v represents valve natural frequency and d_v denotes damping constant. The valve gain k_v in (3.20) is expressed with the form $k_v = Q_N/(i_{max}\sqrt{\Delta p_N/2})$, in which the maximum current is i_{max}, and Q_N, Δp_N are flow and pressure drop at i_{max}.

Finally, the model of an electro-hydraulic actuator is formed by the previous relationships, such as (3.18), (3.19) and (3.20), which can be converted to the form of state-space representation:

$$\dot{x}_{act} = A_{act}x_{act} + B_{act,1}w_{act} + B_{act,2}u_{act}, \tag{3.21}$$

where $x_{act} = \left[\dot{x}_v \; x_v \; p_L \; \dot{x}_c \; x_c\right]^T$ represents state vector, $w_{act} = F_{ext}$ is the disturbance and $u_{act} = i$ is the control input.

3.4.2 Robust Control Design for Actuator Positioning Control

The electro-hydraulic model of the actuator incorporates various parameters. The physical parameters are determined by the system itself, such as β_e depending on the chamber pressure. Other parameter values are provided by identification procedures, for example K_c, K_q in (3.17). In addition, the model used to design the control includes components whose properties are uncertain. The presence of parameter uncertainties in the control design must be involved, to provide an increased performance level. In the synthesis of the $\mathcal{H}_\infty$ control, uncertainty is considered in an unstructured manner, thus the design procedure results in a controller with low performance level. But it is possible to consider uncertainty structure through the complex $\mathcal{H}_\infty/\mu$ method, i.e., through diagonal structure with full or scalar complex blocks. Moreover, it can also be beneficial to express uncertainty by repeated real blocks. Therefore, the method of real $\mathcal{H}_\infty/\mu$ incorporates real and complex blocks for structuring uncertainty, as detailed in Packard and Doyle (1993) and Young et al. (1991).

Here, the $\mathcal{H}_\infty/\mu$ control design procedure is used to ensure system stability, even if some parameter is changed. In the case of the given problem, variations on β_e, K_c and K_q are relevant; see Karpenko and Shapehri (2005). Uncertain parameters can be found in (3.17) and (3.18), which can be considered with their nominal values, together with ranges, which represent their potential variations. The parameters with uncertainty through lower linear fractional transform (LFT) can be formed as

$$K_q = \bar{K}_q(1 + d_q\delta_q) = \mathcal{F}_l\left(\begin{bmatrix} \bar{K}_q & 1 \\ d_q\bar{K}_q & 0 \end{bmatrix}\right) = \mathcal{F}_l(M_q, \delta_q),$$

$$K_c = \bar{K}_c(1 + d_c\delta_c) = \mathcal{F}_l\left(\begin{bmatrix} \bar{K}_c & 1 \\ d_c\bar{K}_c & 0 \end{bmatrix}\right) = \mathcal{F}_l(M_c, \delta_c),$$

$$\beta_e = \bar{\beta}_e(1 + d_e\delta_e) = \mathcal{F}_l\left(\begin{bmatrix} \bar{\beta}_e & 1 \\ d_e\bar{\beta}_e & 0 \end{bmatrix}, \delta_e\right) = \mathcal{F}_l(M_e, \delta_e). \tag{3.22a}$$

In LFT structure, the connection between the output and the input of the block M_e is $\tilde{y}_e = \bar{\beta}_e\tilde{u}_e + u_e$, while δ_e uncertainty block is pulled out of the equation. $\bar{K}_q, \bar{K}_c, \bar{\beta}_e$ stands for the nominal parameter values. The variations of the uncertain parameters are characterized by scalars d_q, d_c, d_e, which indicates the percentage of the permissible variation. Furthermore, the actualdeviations of the parameters are represented by $-1 \leq \delta_q, \delta_c, \delta_e \leq 1$. Through uncertainty formulation, all of the blocks δ_i, $i \in (q, c, e)$ must be pulled out from the equations of motion:

$$\dot{p}_L = \frac{4\bar{\beta}_e}{V_t}\left(\bar{K}_q x_v - \bar{K}_c p_L - A_p \dot{x}_c + c_{l1}\dot{x}_c - c_{l2} p_L\right) + \frac{4}{V_t}\left(\bar{\beta}_e u_q - \bar{\beta}_e u_c + u_e\right), \tag{3.23}$$

where $\tilde{y}_i$ represent specified outputs in (3.17), (3.18). System uncertainties in the state-space representation are handled as disturbances. In this manner, (3.21) is converted as follows:

$$\dot{x}_{ua} = A_{ua} x_{ua} + B_{ua,1} w_{ua} + B_{ua,2} u_{ua}, \tag{3.24}$$

where $x_{ua} = x_{act}$ represents the vector of states, $w_{ua} = \left[F_{ext}\ u_q\ u_c\ u_e\right]^T$ is the disturbance and $u_{ua} = i$ represents the control input.

The aim of the control design for an electro-hydraulic actuator is to provide the positioning of the piston motion x_c with regard to $x_{c,ref}$ reference value, such as

$$z_1 = x_{c,ref} - x_c; \qquad |z_1| \rightarrow min! \tag{3.25}$$

In addition, the system control input must be minimized, which leads to the limitation of i:

$$z_2 = i; \qquad |z_2| \rightarrow min! \tag{3.26}$$

Using z_1 and z_2, performance vector $z_{ua} = \left[z_1\ z_2\right]^T$ can be constructed. Moreover, it is necessary to consider that the controller requires measurement on piston position $y_{ua} = x_c$. The performance and measurement equations are formulated as

$$z_{ua} = C_{ua,1} x_{ua} + D_{ua} u_{ua}, \tag{3.27a}$$

$$y_{ua} = C_{ua,2} x_{ua}. \tag{3.27b}$$

The reference movement $x_{c,ref}$ is defined by a_y, i.e., it depends on the requested control input on the high level. The connection between $x_{c,ref}$ and a_y is formed through a relationship $x_{c,ref} = f(a_y)$, which is illustrated in Fig. 3.3. Its characteristics rely on the suspension construction, especially, the connection of the upper suspension arm and the cylinder. Function f can be determined through measurements, and the results can lead to its nonlinear description.

Through the design of the actuator control, stability of the system can be achieved, together with disturbance rejection against F_{ext}. A further aspect of the control design is that d_q, d_c, d_e are uncertain parameters. The robust $\mathcal{H}_\infty/\mu$ design procedure can have the capability to handle stability issues, and disturbance rejection and parametric uncertainties. The appropriate balancing of the performance criteria is ensured through the scaling of the applied weighting functions. In the design process, $W_{z,1} = \lambda(\varepsilon_1 s + 1)/(T_1 s + 1)$ stands for reducing position error. Proper selection of ε_1, T_1 guarantees reducing position error at low frequencies to $1/\lambda$. Weight $W_{z,2} = 1/i_{max}$ is connected to performance z_2. Δ includes system parametric uncertainties, while W_{ext} and W_{ref} are for scaling disturbances and reference signals. In the design method of $\mathcal{H}_\infty/\mu$ controller, it is requested to find controller K, by which

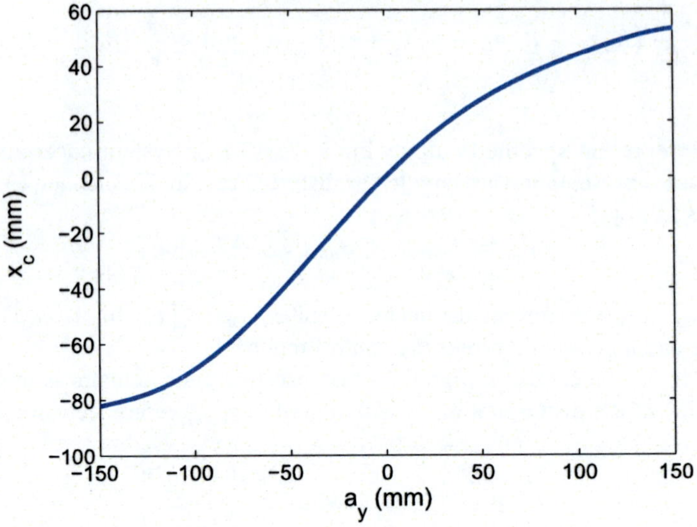

Fig. 3.3 Static map between a_y and x_c

$$\mu_{\tilde{\Delta}}(M(i\omega)) \leq 1, \ \forall\omega \Leftrightarrow \min_{K \in \mathcal{K}_{stab}} \left[\max_{\omega} \mu(M(i\omega))\right], \tag{3.28}$$

where μ is the function of the structured singular value of the system $M(i\omega)$ with a defined uncertainty set $\tilde{\Delta} = diag[\Delta_r, \Delta_m, \Delta_p]$. Δ_r stands for the parametric uncertainties, Δ_m represents unmodeled dynamics and Δ_p describes a fictitious uncertainty block, which involves the performance objectives into the μ framework. It is possible to solve the optimization problem iteratively using a sequential method with the joint minimization of D and K. In step one, D is fixed and the controller K is minimized, which leads to an $\mathcal{H}_\infty$ design. In step two, K is fixed and the value of D is reduced, etc. In this step, the searching of D leads to a convex optimization problem. Although the optimization can have difficulties, ad hoc algorithm, i.e., $D - K$ iteration can be adapted (Packard and Doyle 1993).

3.5 Illustration of the Vehicle Suspension Control Design

The operation of the achieved optimal suspension control on a simulation example is illustrated. In the example, the vehicle cruises on a predetermined path, while the control system aids the chauffeur to ensure path following. For simulation purposes, an E-class vehicle has been chosen, whose weight is 1530 kg, and the power of the engine is 300 kW. In the case of all wheels, the vehicle has double-wishbone type of suspensions, whose construction is independent of each other. In the simulation the design of the control system is carried out through MATLAB®, and the

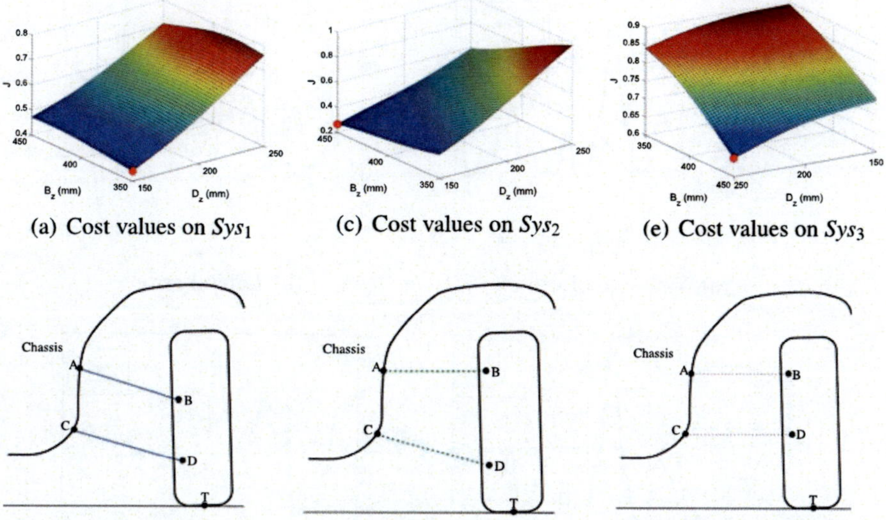

(a) Cost values on Sys_1 (c) Cost values on Sys_2 (e) Cost values on Sys_3

(b) Suspension Sys_1 (d) Suspension Sys_2 (f) Suspension Sys_3

Fig. 3.4 Cost surfaces and geometries

dynamics of the vehicle and the suspension is evaluated under the CarSim® simulation environment.

In the example, three different geometry positions of the arms are analyzed; see the constructions in Fig. 3.4. The presented constructions are the results of the joint optimization of control and construction, with different priorities among the performances.

In the case of the first suspension system, i.e., Sys_1, the aim is to minimize chassis roll value in agreement with the performance of z_2, z_5, which leads to the positions of $((B_z; D_z) = (350; 150))$. Sys_2 denotes the second control and construction, whose role is to reduce change on half-track, and thus, performances z_3, z_6 have priorities. It leads to the geometry values of $((B_z; D_z) = (450; 150))$. The third setup, i.e., Sys_3 system operates the control input as stated in the minimization of performance z_1 $((B_z; D_z) = (450; 250))$. The connection of construction, performance weights and the resulting cost surface $\mathcal{J}(B_z, D_z)$ is illustrated in Fig. 3.4a–c. The (3.14) optimization method leads to different optimums for controllers and suspension constructions, depending on the weight selection strategy.

In Fig. 3.5, the achieved suspension control and constructions are evaluated through the cruising on a section of the Waterford Michigan Race Track. Figure 3.5a describes the tracking of yaw rate using the three simultaneously designed control systems. Although an uncontrolled vehicle cannot ensure trajectory tracking and causes hazardous situations, all of the vehicles with the proposed suspension control are suitable for following the path, i.e., minimizing yaw rate error. Moreover, variable-geometry vehicle suspension reduces required driver steering

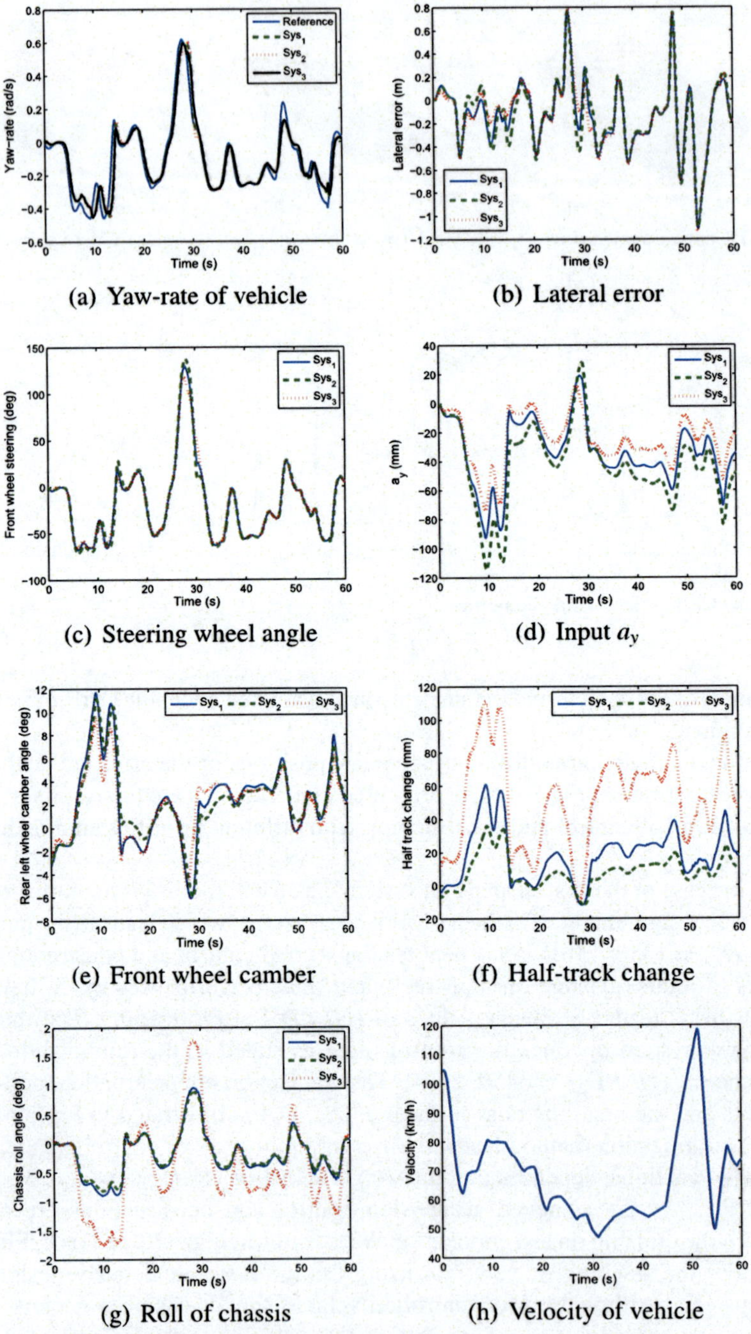

(a) Yaw-rate of vehicle (b) Lateral error

(c) Steering wheel angle (d) Input a_y

(e) Front wheel camber (f) Half-track change

(g) Roll of chassis (h) Velocity of vehicle

Fig. 3.5 Simulation with vehicle suspension control (©2013 IEEE. Reprinted, with permission, from Németh and Gáspár 2013)

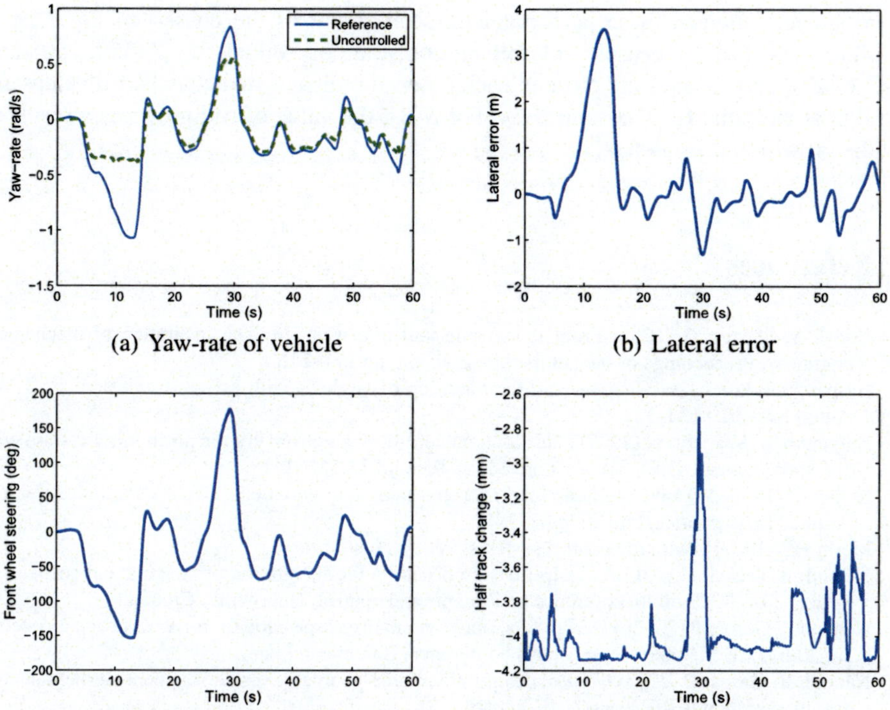

(a) Yaw-rate of vehicle (b) Lateral error

(c) Steering wheel angle (d) Half-track change

Fig. 3.6 Simulation without suspension control

intervention, as depicted in Fig. 3.5c. Compare these maximum steering wheel values with Fig. 3.6c, where the motion of the vehicle without control intervention is simulated. Thus, the proposed suspension system is also capable to ensure trajectory tracking, and to facilitate driver steering.

Figure 3.5d describes a_y intervention for each suspension. In the case of Sys_3, the reduction of a_y is favored, whereas in other cases increased control intervention during the maneuver is generated. The actuation value a_y is significantly influenced by the selection of $W_{p,i}$. Wheel tilting on front wheels (Fig. 3.5e) are close to each other, which leads to the following consequences. Each variable-geometry vehicle suspension ensures path following, i.e., the tracking performance level of each control system is high. On the other hand, using different control inputs a_y can develop almost approximately the same tilting, which is a constructional consequence. These appointments are the outcomes of the simultaneous design method of construction and control. Figure 3.5f shows the change of half-track on the controlled front wheels. In the case of Sys_2 system, its value can be significantly decreased.

As a summary, a minimum on a_y and on ΔB with different $(B_z; D_z)$ can be achieved. Though, through Sys_2, the values of half-track change can be reduced, it leads to the biggest control input values. Nevertheless, Sys_1 is able to guarantee a balance between these performances. Roll signal of the chassis in Fig. 3.5g is illustrated. From the aspect of roll motion, the cost values of $\mathcal{J}(350, 150)$ and $\mathcal{J}(450, 150)$ at Sys_1 are close to each other. It indicates that chassis roll values are almost the same for Sys_1 and Sys_2. For Sys_3, the value of roll grows as a result of the growth in $|h_{CG} - h_{M,st}|$.

References

Gough V, Shearer G (1956) Front suspension and tyre wear. In: The institution of mechanical engineers, Proceedings of the automobile division, pp 171–216

Gáspár P, Bokor J (2005) Improving the handling of vehicles with active suspension. Int J Veh Auton Syst 3(2):134–151

Karpenko M, Shapehri N (2005) Fault-tolerant control of a servohydraulic positioning system with crossport leakage. IEEE Trans Control Syst Technol 13:155–161

Kelley C (1995) Iterative methods for linear and nonlinear equations. Society for Industrial and Applied Mathematics, Philadelphia USA

Meritt HE (1967) Hydraulic control systems. Wiley, New York

Németh B, Gáspár P (2012a) Control design based on the integration of steering and suspension systems. In: IEEE multiconference on systems and control, Dubrovnik, Croatia

Németh B, Gáspár P (2012b) Design of variable-geometry suspension for driver assistance systems. In: 20th IEEE Mediterranean control conference, Barcelona, Spain

Németh B, Gáspár P (2013) Control design of variable-geometry suspension considering the construction system. IEEE Trans Veh Technol

Packard A, Doyle J (1993) The complex structured singular value. Automatica 29(1):71–109

Šulc B, Jan JA (2002) Non linear modelling and control of hydraulic actuators. Acta Polytech 42(3):41–47

Wijnheijmer F, Naus G, Post W, Steinbuch M, Teerhuis P (2006) Modelling and LPV control of an electro-hydraulic servo system. In: IEEE computer aided control system design, Münich

Young P, Newlin M, Doyle J (1991) μ analysis with real parametric uncertainty. In: Proceeding of the conference on decision and control, Brighton, England, pp 1251–1256

Chapter 4
SOS-Based Modeling, Analysis and Control

4.1 Motivations

In various vehicle control problems, the modeling of vehicle dynamics in linear form can be adequate to create control systems. Nevertheless, the understanding of operation in detail can require nonlinear methods. Especially, in the case of variable-geometry vehicle suspensions, the nonlinear characteristics of lateral tire forces request enhanced synthesis and analysis methods. In this chapter, a system analysis method based on the Sum-of-Squares (SOS) programming technique is applied, which is able to involve polynomial formulation in vehicle modeling. The focus of this chapter is to approximate Controlled Invariant Sets, which provide information on the effectiveness of the control intervention, i.e., on the maximum of the control forces for each wheel.

A motivational example of the necessity of SOS-based analysis is presented in Fig. 4.1. It shows the result of a linear approximation of reachability sets, regarding the lateral dynamics using variable-geometry vehicle suspension (Németh and Gáspár 2012). The analysis studies two scenarios: the reachable side slip values on each axle (α_1, α_2), which are calculated using limited tilting and steering intervention with different configurations. In the case of the first configuration, only steering control is applied, but, in the second case, an integrated intervention of tilting and steering is used. It is well demonstrated that the results of the scenarios are close to each other, i.e., the impact of wheel tilting is low. Despite the reachable set analysis of Németh and Gáspár (2013a), the results suggest that modifying the angle of the wheel camber angle can have great benefit in vehicle control. Hence, in this example linear analysis does not adequately handle the effectiveness of variable-geometry suspension in vehicle dynamics. The reason for this contradiction is in the modeling of the vehicle dynamics. A further consecutive disadvantage of linear analysis is that nonlinear characteristics of the tire are neglected, which influences the stability of

All figures are reprinted from Németh and Gáspár (2017), with permission from Elsevier.

B. Németh and P. Gáspár, *Control of Variable-Geometry Vehicle Suspensions*, Advances in Industrial Control, https://doi.org/10.1007/978-3-031-30537-5_4

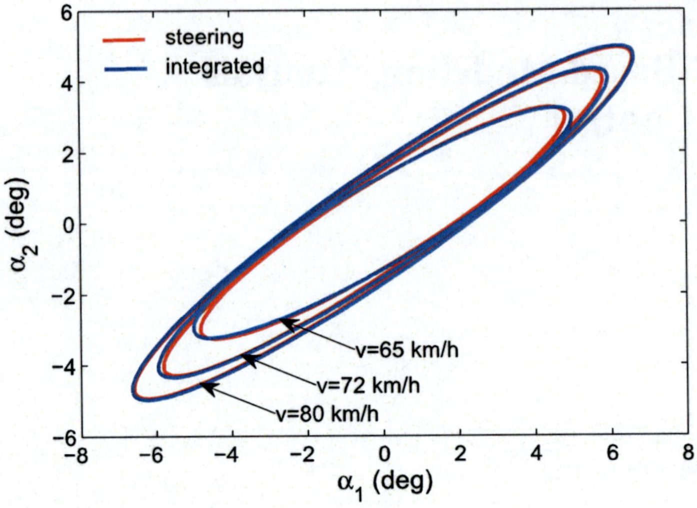

Fig. 4.1 Reachable sets of the systems

the vehicle. In the case of a linear description of lateral vehicle dynamics $\dot{x} = Ax + B_1 w + B_2 u$, the eigenvalues of A matrix provide information on vehicle stability. Even though the eigenvalues of A are influenced by longitudinal velocity v, instability of the system cannot be achieved by increasing v within its conventional range. In addition, the stability of the system is significantly affected by the characteristics of the nonlinear tire. The analysis shown in Fig. 4.1 implies that reachable sets have an increasing tendency at increasing v. Nevertheless, (Pacejka 2004) depicts that in the case of lateral dynamics the increasing v lead to stability margin reduction. Therefore, stability and performance problems at high-velocity values cannot be effectively examined by linearly reachable set calculation.

4.2 Analysis-Oriented Formulation of Nonlinear Lateral Vehicle Dynamics

There are different effects on the wheels induced by the variable-geometry suspension system: variation of camber and steering angles. The connection between these effects is defined by suspension construction. The nonlinear vehicle model in this chapter is developed for the example of the double-wishbone type of variable-geometry suspension system. Nevertheless, the method can be applied to another type of suspension construction.

4.2.1 Formulation of Nonlinear Lateral Model

Formulation of tire model, i.e., lateral forces is a crucial problem in the design of vehicle control systems; some examples of the existing models can be found in the works of Pacejka (2004), Kiencke and Nielsen (2000) and de Wit et al. (1995). Here, lateral tire force characteristics are approximated by applying polynomial functions. The benefit of this model is its composition, which can be used effectively in analytic techniques, e.g., finding Lyapunov functions for proving stability or controllability. In addition, the polynomial functions are properly matched to the characteristics of the lateral tire force, e.g., Hirano et al. (1993), Sadri and Wu (2013) and López et al. (2014) motivate polynomial formulation of tire force characteristics.

Figure 4.2 shows the scheme of the vehicle model, whose dynamics can be formed through the following relationships: begin subequations

$$J\ddot{\psi} = F_{lat,1}(\alpha_1)l_1 - F_{lat,2}(\alpha_2)l_2 = F_1(\alpha_1)l_1 - F_2(\alpha_2)l_2 + G(\alpha_1)l_1\gamma, \tag{4.1a}$$

$$mv\left(\dot{\psi} + \dot{\beta}\right) = F_{lat,1}(\alpha_1) + F_{lat,2}(\alpha_2) = F_1(\alpha_1) + F_2(\alpha_2) + G(\alpha_1)\gamma, \tag{4.1b}$$

in which relationships l_1 and l_2 are vehicle parameters, J represents yaw inertia and m is the mass of the vehicle. Moreover, β is side slip, $\dot{\psi}$ denotes yaw rate and δ, γ are wheel steering and tilting angles. Lateral forces are denoted by functions $F_{lat,1}(\alpha_1)$, $F_{lat,2}(\alpha_2)$, in which α_1, α_2 are side slip angles on the axles. The following relationships between the signals can also be formed as $\tan(\alpha_2) = (l_2\dot{\psi} - v\sin\beta)/(v\cos\beta)$, $\tan(\delta - \alpha_1) = (l_1\dot{\psi} + v\sin\beta)/(v\cos\beta)$. Under conventional cruising, α_i is usually under 10°, which results in the possibility of approximations $cos\beta \approx 1$, $\sin\beta \approx \beta$.

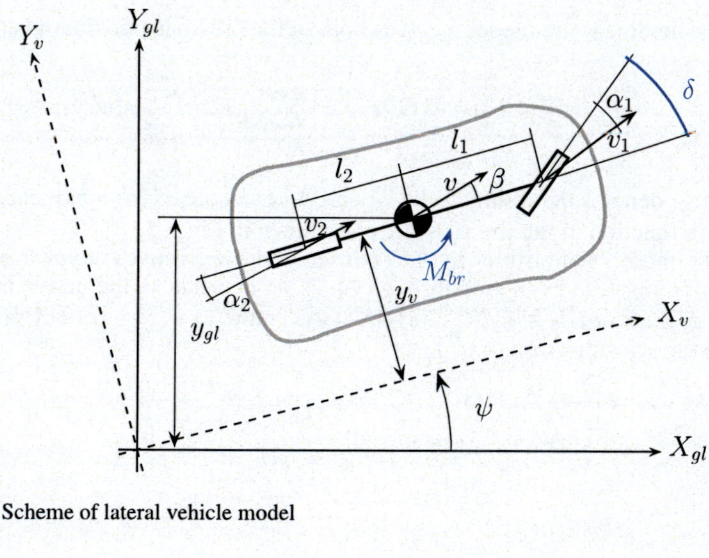

Fig. 4.2 Scheme of lateral vehicle model

This simplification leads to a slight relative error under normal road-vehicle conditions, such as less than 1%. Therefore, the α_1, α_2 can be approximated as follows:

$$\alpha_1 = \delta - \beta - \frac{\dot{\psi} l_1}{v}, \tag{4.2a}$$

$$\alpha_2 = -\beta + \frac{\dot{\psi} l_2}{v}. \tag{4.2b}$$

From these terms, β and $\dot{\psi}$ can be expressed as

$$\dot{\psi} = v \frac{\alpha_2 - \alpha_1 + \delta}{l_1 + l_2}, \tag{4.3a}$$

$$\beta = -\frac{\alpha_1 l_2 + \alpha_2 l_1 - l_2 \delta}{l_1 + l_2}. \tag{4.3b}$$

In the operation range of the variable-geometry vehicle suspension, the fitting of the tire characteristics in the nonlinear range above can be carried out as follows.

- Function $F(\alpha)$ relies upon α nonlinearly: $F(\alpha) = \sum_{j=1}^{n} c_j \alpha^j$. In many control applications the lateral forces can be approached by linear functions, which results in a simple formulation. Nevertheless, that model only in narrow tire side slip region can be used.
- The proposed polynomial model of lateral tire force can be extended with the impact of wheel tilting $G(\alpha)$ as $G(\alpha) = \sum_{k=0}^{m} g_k \alpha^k$. Accordingly, the effectiveness of the actuator intervention is affected by α_i.

Thus, the nonlinear tire model involves both of the previous polynomial approximations:

$$F_{lat}(\alpha) = F(\alpha) + G(\alpha)\gamma = \sum_{j=1}^{n} c_j \alpha^j + \sum_{k=0}^{m} g_k \alpha^k \gamma, \tag{4.4}$$

in which γ denotes the tilting of the wheel. An example of the nonlinear characteristics as a function of the tire side slip α is shown in Fig. 4.3.

Furthermore, relationships in (4.1) consist of the derivatives of yaw rate and side slip, which must be expressed through (4.3). At constant v, the derivatives can be formed as $\dot{\beta} = -(\dot{\alpha}_1 l_2 + \dot{\alpha}_2 l_1 - l_2 \dot{\delta})/(l_1 + l_2)$ and $\ddot{\psi} = v(\dot{\alpha}_2 - \dot{\alpha}_1 + \dot{\delta})/(l_1 + l_2)$. It leads to the reformulation of (4.1) as

$$\dot{\alpha}_2 - \dot{\alpha}_1 = \left[\frac{l_1 + l_2}{J v} (F_1(\alpha_1) l_1 - F_2(\alpha_2) l_2) \right] - \dot{\delta} + \frac{l_1 (l_1 + l_2)}{J v} G(\alpha_1)\gamma, \tag{4.5a}$$

$$\dot{\alpha}_1 l_2 + \dot{\alpha}_2 l_1 = v(\alpha_2 - \alpha_1) - \frac{l_1 + l_2}{m v} [F_1(\alpha_1) + F_2(\alpha_2)] + v\delta + l_2 \dot{\delta} - \frac{l_1 + l_2}{m v} G(\alpha_1)\gamma. \tag{4.5b}$$

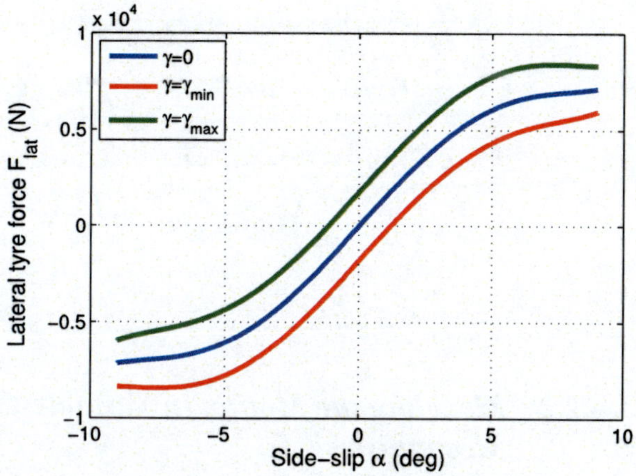

Fig. 4.3 Modeling of lateral tire force F_{lat}

Next, these terms are used to convert (4.1) into state-space representation with polynomial functions $\dot{x} = f(x) + gu$, in which f, g are matrices, $x = \begin{bmatrix}\alpha_1 \ \alpha_2\end{bmatrix}^T$ represents state vector and u denotes control intervention. In the novel model, the lateral tire force nonlinearities are hidden in F_1, F_2 and G. Nonetheless, (4.5) includes the time derivative of the steering angle. During the analysis of the control intervention on the vehicle motion, $|\delta|$ is limited, and the proposed vehicle model must have accuracy in this operating range. Therefore, the following approximations are used for the application of the actuation limits:

$$\max(|\dot{\delta}|) = \max\left(\frac{|\dot{\delta}|}{|\delta|}\right) \cdot \max(|\delta|) = \xi \cdot \max(|\delta|), \tag{4.6a}$$

$$\dot{\delta} \approx \xi \cdot \delta, \tag{4.6b}$$

where parameter ξ indicates the connection between the altering speed of δ and the maximum steering value. As $\max \delta$ is a specific fixed constraint in the analysis of the actuator, a high-value of ξ represents a rapidly altering steering signal, while a slowly altering steering signal is modeled with a low ξ.

Finally, the system is represented by a polynomial state space using (4.5), and the replacement of (4.6) is given as follows:

$$\dot{x} = \begin{bmatrix}\dot{\alpha}_1 \\ \dot{\alpha}_2\end{bmatrix} = \begin{bmatrix}\mathrm{f}_1(\alpha_1, \alpha_2) \\ \mathrm{f}_2(\alpha_1, \alpha_2)\end{bmatrix} + \begin{bmatrix}h_{11} \ h_{12} \\ h_{21} \ h_{22}\end{bmatrix}\begin{bmatrix}\delta \\ \gamma\end{bmatrix}, \tag{4.7}$$

where

$$\mathrm{f}_1 = \frac{l_1}{Jv}\left[F_2(\alpha_2)l_2 - F_1(\alpha_1)l_1\right] + \frac{v}{l_1+l_2}(\alpha_2-\alpha_1) - \frac{1}{mv}\left[F_1(\alpha_1)+F_2(\alpha_2)\right],$$
$$\mathrm{f}_2 = \frac{l_2}{Jv}\left[F_1(\alpha_1)l_1 - F_2(\alpha_2)l_2\right] + \frac{v}{l_1+l_2}(\alpha_2-\alpha_1) - \frac{1}{mv}\left[F_1(\alpha_1)+F_2(\alpha_2)\right],$$
$$h_{11} = \frac{v}{l_1+l_2} + \xi, \quad h_{21} = \frac{v}{l_1+l_2},$$
$$h_{12} = -\left(\frac{l_1^2}{Jv} + \frac{1}{mv}\right)G(\alpha_1), \quad h_{22} = \left(\frac{l_1 l_2}{Jv} - \frac{1}{mv}\right)G(\alpha_1).$$

4.2.2 Modeling the Motion in Variable-Geometry Suspension Mechanism

The connection between steering and tilting angles in the rest is introduced as follows. The scheme of variable-geometry vehicle suspension construction in Fig. 4.4 is illustrated. Intervention a_y leads to the front wheel rotation around axis KD, i.e., γ and δ are modified at the same time. In this modification, the vertical position of K has an important role, for example, the distribution of γ and δ depends on it. Hence, K_z must be selected on the way, which results in the highest impact on the lateral tire forces. A method on determining the connections of $\gamma = \mathfrak{f}_\gamma(a_y, K_z)$ and $\delta = \mathfrak{f}_\delta(a_y, K_z)$ can be found in Németh and Gáspár (2013b).

In the given model of the suspension, arms and bodies of the structure are components linked by joints to the chassis of the vehicle. Figure 4.5a, b illustrates the characteristics of γ, δ for K_z with different values. The example demonstrates that the selection of K_z can have a high impact on the value of δ, and, similarly, it slightly

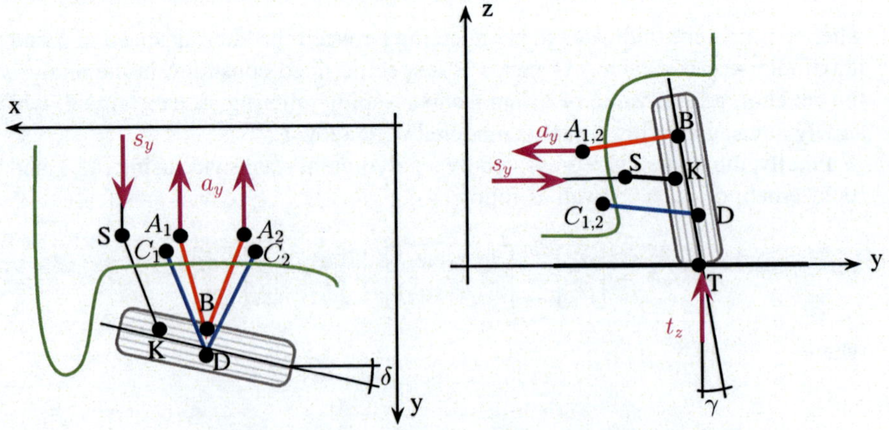

(a) Up view (b) Front view

Fig. 4.4 Illustration of camber and steering angles related to wheel position

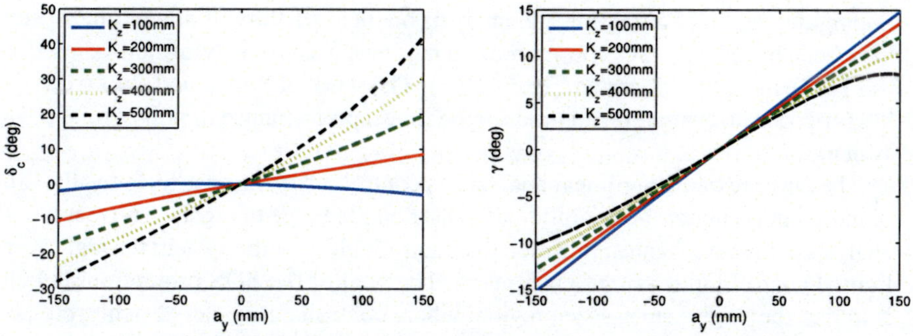

(a) δ change (b) γ change

Fig. 4.5 Influence of K_z on the relationship between δ and γ

influences γ. *KD* is the axis of wheel rotation during actuation a_y, thus its orientation affects the connection between the angles. Since δ and γ are usually in conflict, the appropriate solution for the parameter K_z must be found. In the examined construction, K_z has a great effect on δ, and by increasing K_z, a high lateral tire force can be achieved. As a consequence, the steering angle and camber angle are related to the actuation, i.e., $\delta = \mathfrak{f}_\delta(a_y)$ and $\gamma = \mathfrak{f}_\gamma(a_y)$. This illustrated that the construction parameter has a great effect on the operation of the variable-geometry suspension system. It is shown by the example that a trade-off between the parameters must be found. A selection of K_z with high-value can be beneficial, because δ in a wider range can be achieved through a_y, but it is disadvantageous on the value of γ.

4.3 Analysis of Actuation Efficiency Through Nonlinear Method

Basic concepts of actuation efficiency examination through Sum-of-Squares (SOS) programming in this section are presented, focusing on the systems with polynomial dynamics. First, the theoretical background is presented and, second, its application on variable-geometry vehicle suspension is shown.

Many studies focus on SOS programming which has been developed for control purposes over the past decade. Substantial theorems according to SOS programming have been presented and discussed by Parrilo (2003). The connections between conditions in SOS programming and Linear Matrix Inequalities (LMI), i.e., transformation methods and solution on the polynomial analysis problem, have been found in Prajna et al. (2004). SOS programming has been applied for various control problems, such as approximation of reachability sets or design of control systems, as it has been presented by Jarvis-Wloszek et al. (2003). The approximation method of the maximum controlled invariant set for controlled polynomial systems has been provided by Korda et al. (2013). Moreover, in Tan and Packard (2008) iteration process for finding the region of attraction and analysis on local stability of polynomial systems have been introduced. SOS programming has been used for solving non-convex control problems; see Scherer and Hol (2006). The work of Topcu and Packard (2009) has proposed an evaluation method for the robust performance of polynomial control systems.

The purpose of the nonlinear analysis on control intervention is to define the limits of intervention under the condition of peak-bounded type of signal. The result of the analysis is the set of states, in which the controllability of the system through a given limited control input can be guaranteed. The goal of the SOS programming below is to find the largest state-space region, where controllability can be achieved, i.e., a method on the calculation of Controlled Invariant Sets is formed.

4.3.1 Method of Computation for Controlled Invariant Sets

The following definitions are crucial for understanding the aim of the SOS programming method, i.e., definitions of polynomial and SOS (Jarvis-Wloszek et al. 2003):

Definition 4.1 A **Polynomial** f in n variables is a finite linear combination of the functions $m_\alpha(x) := x^\alpha = x_1^{\alpha_1} x_2^{\alpha_2} \cdots x_n^{\alpha_n}$ for $\alpha \in \mathbb{Z}_+^n$, $\deg m_\alpha = \sum_{i=1}^n \alpha_i$:

$$f := \sum_\alpha c_\alpha m_\alpha = \sum_\alpha c_\alpha x^\alpha \tag{4.8}$$

with $c_\alpha \in \mathbb{R}$. Define $\mathbb{R}$ to be the set of all polynomials in n variables. The degree of f is determined as $f := \max_\alpha \deg m_\alpha$.

Definition 4.2 The set of **Sum-of-Squares (SOS)** polynomials in n variables is defined as

$$\Sigma_n := \left\{ p \in \mathscr{R}_n \;\middle|\; p = \sum_{i=1}^{t} f_i^2,\ f_i \in \mathscr{R}_n,\ i = 1, \ldots, t \right\}. \tag{4.9}$$

The polynomial state-space representation of the system is defined as follows, see (4.7):

$$\dot{x} = f(x) + gu, \tag{4.10}$$

in which matrix $f(x)$ smooth polynomial functions are included with the initial value $f(0) = 0$. During the SOS-based analysis, the system is considered to have single input.

Control Lyapunov Function is used to describe global asymptotic stability of $\dot{x} = f(x) + gu$, which is determined as follows (Sontag 1989):

Definition 4.3 A smooth, proper and positive-definite function $V : \mathbb{R}^n \to \mathbb{R}$ is a Control Lyapunov Function for a system if

$$\inf_{u \in \mathbb{R}} \left\{ \frac{\partial V}{\partial x} f(x) + \frac{\partial V}{\partial x} g \cdot u \right\} < 0 \tag{4.11}$$

for each $x \neq 0$.

In accordance with Definition 4.3, two cases are distinguished:

- if $\frac{\partial V}{\partial x} f(x) < 0$ then the system is stable and $u \equiv 0$. This stability scenario is contained by the next two stability criteria.
- relation $\frac{\partial V}{\partial x} f(x) > 0$ is related to the unstable states of the system. Nonetheless, its stabilization is achievable:
  - u with upper peak-bound value is able to stabilize the unstable system if

$$\frac{\partial V}{\partial x} g < 0 \quad \text{and} \quad \frac{\partial V}{\partial x} f(x) + \frac{\partial V}{\partial x} g \cdot u_{max} < 0. \tag{4.12}$$

  - Moreover, u with its lower peak-bound is able to stabilize the unstable system if

$$\frac{\partial V}{\partial x} g > 0 \quad \text{and} \quad \frac{\partial V}{\partial x} f(x) - \frac{\partial V}{\partial x} g \cdot u_{max} < 0. \tag{4.13}$$

  In the previous inequalities, the equivalence of the peak-bound amplitudes is considered, i.e., $u_{min} = -u_{max}$.

The aim of the Control Lyapunov Function is to approximate the maximum Controlled Invariant Set of the system, which is the $V(x) = 1$ level set of the Control Lyapunov Function. Hence, the former criteria on stability have to be guaranteed for sets described by $V(x) \leq 1$. Although the formulation of finding the maximum Controlled Invariant Set can be easy, its computation in practice may be difficult. Therefore, a three-step procedure for the approximation is recommended below.

Step 1: The region of attraction of the uncontrolled system $\dot{x} = f(x)$ is defined as an initial set. In this step, the maximum level set of $V_0 = 1$ is determined, which is included in the stable region. The SOS-based calculation of the attraction region is described in the work of Jarvis-Wloszek (2003).

Step 2: For the maximization process of the Controlled Invariant Set, a parameter η is selected. It is used for scaling the invariant set, i.e., $V_\eta = V_0 \cdot \eta$ is examined as a Local Control Lyapunov Function. It results in a candidate Controlled Invariant Set S_η through $V_\eta = 1$. The question is whether the peak-bounded finite u is able to stabilize the system in S_η. Depending on parameter η, the candidate level set can be scaled, such as increased or decreased. The computation method on the Local Control Lyapunov using the SOS tools can be found in Tan and Packard (2008).

Step 3: Finally, the candidate set of S_η is evaluated from the aspects of acceptability and also of increasing possibility. During the evaluation, $u_{min} = -u_{max}$ and

u_{max} peaks in the intervention are involved. The set $S_{inst} = \left\{ x \in \mathscr{R}_n \middle| \frac{\partial V}{\partial x} f(x) > 0 \right\}$ is defined as the region with unstable states. Moreover, $S_{min} = \left\{ x \in \mathscr{R}_n \middle| \frac{\partial V}{\partial x} f(x) - \frac{\partial V}{\partial x} g \cdot u_{max} > 0 \right\}$ reflects on the states, which cannot be stabilized by u_{min}. Correspondingly, $S_{max} = \left\{ x \in \mathscr{R}_n \middle| \frac{\partial V}{\partial x} f(x) + \frac{\partial V}{\partial x} g \cdot u_{max} > 0 \right\}$ is the region, in which the states through u_{max} cannot be stabilized. If S_η is an appropriate Controlled Invariant Set, then

$$S_\eta \bigcap S_{inst} \bigcap S_{min} \bigcap S_{\max} = \emptyset. \tag{4.14}$$

This advantage of the emptiness condition on the intersection of the sets is that it can be examined through simple tools through the plot of S_{inst}, $S_{\max}$, S_{min} and S_η. Furthermore, in the case of acceptable S_η, the value of η can be increased in the next step to finding the maximum Controlled Invariant Set.

4.3.2 *Illustration of the Effectiveness of the Intervention*

Next, an illustration of the analysis of Maximum Controlled Invariant Set S is carried out. The examination requires a polynomial tire model, which is defined as

$$F_1(\alpha_1) = \sum_{j=1}^{6} c_{j,1}\alpha_1^j, \quad F_2(\alpha_2) = \sum_{j=1}^{6} c_{j,2}\alpha_2^j, \quad G(\alpha_1) = g_0 + g_4\alpha_1^4; \tag{4.15}$$

see (4.4). The analysis also examines the effect of K_z value on the computed Controlled Invariant Sets. It is considered that the maximum of variable-geometry vehicle suspension control input is $|a_{y,max}| = 150$ mm; see Fig. 4.5. In the example, the impact of K_z selection through steering and wheel tilting on the vehicle control is analyzed. It is compared with a front wheel steering system, which has $|\delta_{max}|$ steering intervention constraint, where $|\delta_{max}|$. Thus, through the analysis the additional impact of wheel tilting on the vehicle dynamics can be illustrated.

Maximum Controlled Invariant Sets for given K_z values at different vehicle speeds can be seen in Fig. 4.6. In the case of $K_z = 100$ mm selection, wheel tilting has a slight impact on the region of stable states. However, increasing K_z to 300 mm induces increasing impact on stable states, i.e., joint steering and tilting intervention is preferred. Consequently, simultaneous operation of the steering wheel and wheel tilt results in an increase of S, see scenarios $K_z = 300$ mm, $K_z = 500$ mm, though this growth has higher relevance at $K_z = 300$ mm. Nevertheless, K_z is requested to limit, because $K_z = 500$ mm results in the reduction of the set. Its reason is that the maximum of γ has reduced value for high K_z. The illustration shows that the selection of K_z can greatly affect the size of the Maximum Controlled Invariant Set,

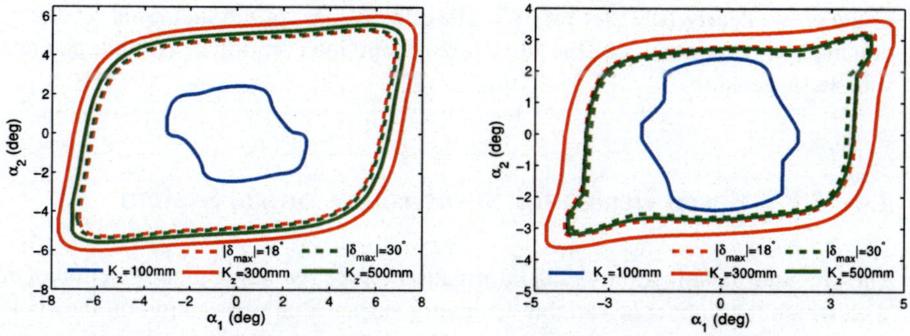

(a) $v = 72km/h$ (b) $v = 80km/h$

Fig. 4.6 Maximum controlled invariant sets

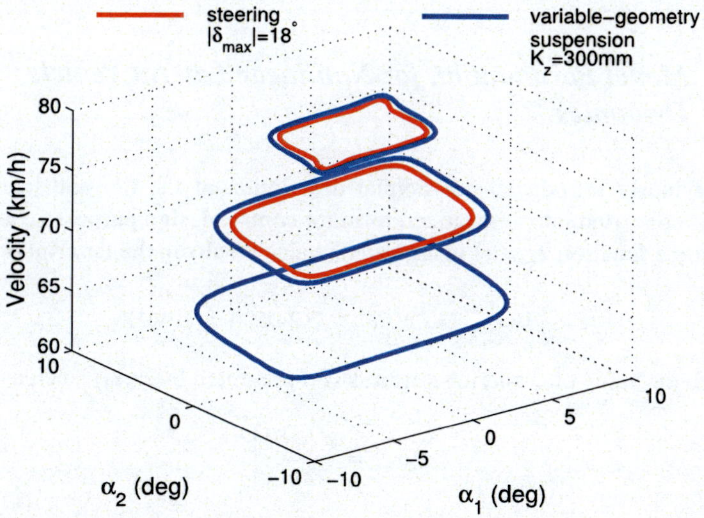

Fig. 4.7 Maximum controlled invariant sets of the systems

i.e., it is necessary to find K_z, which results in the joint increase in the impact of tilting and steering.

In the example introduced, the best choice is selecting $K_z = 300$ mm. Figure 4.7 shows the illustration of Maximum Controlled Invariant Sets for different longitudinal vehicle speeds. The illustration shows that at higher speed values, the set of unstable states grows. It results in the reduction of the controllable region with limited u. Since at 65 km/h speed the validity of the suspension and vehicle models is reached, i.e., the impact of wheel camber angle at $v > 65$ km/h is relevant, when steering and tilting operate simultaneously, S regions become larger, compared to the intervention of pure steering. Consequently, the distance of the set boundary from the origin grows nearly 10%; see Fig. 4.7. The effectiveness of the suggested set calculation method is apparent, i.e., the advantage of joint intervention in variable-geometry vehicle suspension.

4.4 LPV-Based Design for Suspension Control System

The previous examination yields information on the recommended selection of K_z, with which the impact of variable-geometry suspension intervention on the vehicle dynamics can be maximized. Since the system influences camber and steering angles simultaneously, the design of suspension control through the modification of a_y can be challenging.

4.4.1 *Model Formulation for Nonlinear Lateral Vehicle Dynamics*

The previously introduced polynomial description of the tire and lateral vehicle model is converted into a linearized form for control design purposes. Linearization of tire force function $F_{lat}(\alpha)$ in a given α_0 value results in the description

$$F(\alpha)\big|_{\alpha_0} = F_0(\alpha_0) + c(\alpha_0)\alpha + G_0(\alpha_0)\gamma, \tag{4.16}$$

in which the value of cornering stiffness is represented by $c(\alpha_0)$, such as

$$c(\alpha_0) = \left.\frac{dF(\alpha)}{d\alpha}\right|_{\alpha_0} \tag{4.17}$$

stands for the linear slope at α_0. Moreover, $F_0(\alpha_0), c(\alpha_0)$ parameters rely upon α_0. The values of these parameters can be derived from the polynomial tire force model, see (4.4), i.e., $F_0(\alpha_0) = F(\alpha_0)$ and $G_0(\alpha_0) = G(\alpha_0)$. Accordingly, the original nonlinear polynomial model (4.4) is converted into a linear α_0-dependent form. An illustration of the linearization method of the tire force can be found in Fig. 4.8.

Vehicle dynamics in a lateral direction through the following relations can be formulated:

$$J\ddot{\psi} = F_1(\alpha_1)\big|_{\alpha_{0,1}} l_1 - F_2(\alpha_2)\big|_{\alpha_{0,2}} l_2 + G_0(\alpha_{0,1}) l_1 \gamma, \tag{4.18a}$$

$$mv\left(\dot{\psi} + \dot{\beta}\right) = F_1(\alpha_1)\big|_{\alpha_{0,1}} + F_2(\alpha_2)\big|_{\alpha_{0,2}} + G_0(\alpha_{0,1})\gamma, \tag{4.18b}$$

in which vehicle mass is represented by m, J denotes yaw inertia and l_1, l_2 are geometrical vehicle parameters. $\dot{\psi}$ is yaw rate signal and side-slip angle is represented

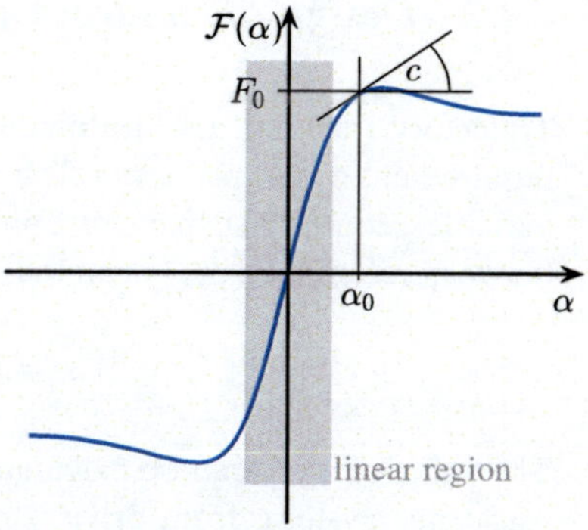

Fig. 4.8 Modeling of lateral tire force

by β. Moreover, $F_1(\alpha_1)\big|_{\alpha_{0,1}}$ and $F_2(\alpha_2)\big|_{\alpha_{0,2}}$ stand for the lateral tire forces, which are around $\alpha_{0,1}, \alpha_{0,2}$ linearized.

The combination of lateral vehicle model (4.18) and tire force characteristics (4.16) leads to the following dynamic relations:

$$J\ddot{\psi} = \big(F_{0,1}(\alpha_{0,1})l_1 - F_{0,2}(\alpha_{0,2})l_2\big) + G_0(\alpha_{0,1})l_1\gamma + c_1(\alpha_{0,1})l_1\left(\delta - \beta - \frac{\dot{\psi}l_1}{v}\right) - c_2(\alpha_{0,2})l_2\left(-\beta + \frac{\dot{\psi}l_2}{v}\right), \quad (4.19a)$$

$$mv\left(\dot{\psi} + \dot{\beta}\right) = \big(F_{0,1}(\alpha_{0,1}) + F_{0,2}(\alpha_{0,2})\big) + G_0(\alpha_{0,1})\gamma + c_1(\alpha_{0,1})\left(\delta - \beta - \frac{\dot{\psi}l_1}{v}\right) - c_2(\alpha_{0,2})\left(-\beta + \frac{\dot{\psi}l_2}{v}\right). \quad (4.19b)$$

In the control synthesis process, it is necessary to consider the physical intervention of the variable-geometry vehicle suspension, i.e., signal a_y instead of steering and tilting actuations is used. Linear connections between a_y and steering and tilting are formed by the functions $\gamma\ \delta = \mathfrak{f}_\delta(a_y)$, $\gamma = \mathfrak{f}_\gamma(a_y)$; see Németh and Gáspár (2013b). Incorporating $\mathfrak{f}_\delta, \mathfrak{f}_\gamma$ in (4.19), the following state-space representation is yielded:

$$\dot{x} = A(\alpha_{0,1}, \alpha_{0,2})x + B_1(\alpha_{0,1}, \alpha_{0,2})w + B_2(\alpha_{0,1}, \alpha_{0,2})u + W(\alpha_{0,1}, \alpha_{0,2}), \quad (4.20)$$

in which x vector of states consists of $x = \left[\dot{\psi}\ \beta\right]^T$, and $u = a_y$ control intervention; disturbance w are formed. The signal of w is the disturbance. Furthermore, static disturbance W with constant values is formed as

$$W = \begin{bmatrix} \frac{F_{0,1}(\alpha_{0,1})l_1 - F_{0,2}(\alpha_{0,2})l_2}{J} \\ \frac{F_{0,1}(\alpha_{0,1}) + F_{0,2}(\alpha_{0,2})}{mv} \end{bmatrix}. \quad (4.21)$$

The proposed description (4.20) is a LPV-based control-oriented formulation on the lateral dynamics of the vehicle, whose scheduling parameters are $\alpha_{0,1}, \alpha_{0,2}$.

4.4.2 Design of Control via LPV-Based Method

Trajectory tracking, i.e., following a reference path or yaw rate, is a performance requirement with a primary role against the controlled vehicle system. In the task of variable-geometry suspension control design, the purpose is to reduce the deviation between reference ($\dot{\psi}_{ref}$) and the real yaw rate signals:

$$z = |\dot{\psi}_{ref} - \dot{\psi}| \rightarrow min. \tag{4.22}$$

The value of $\dot{\psi}_{ref}$ can be calculated from road curvature or, in the case of a driver assistance system, from driver steering intervention δ_d. For instance, $\dot{\psi}_{ref}$ can be computed through a dynamic system, as described in Rajamani (2005):

$$\dot{\psi}_{ref} = \frac{v}{d} \cdot (1 - e^{-\frac{t}{\tau}}) \cdot \delta_d,$$

in which $d = l_1 + l_2 + \eta v^2/g$ is determined by geometrical parameters and velocity of the vehicle, τ is the time constant, g is the gravitational coefficient and η represents the understeer gradient.

Synthesis of the controller on the formed parameter-dependent system representation (4.20) is based, which has the following scheduling variables:

$$\rho_1 = \alpha_{0,1}, \tag{4.23a}$$

$$\rho_2 = \alpha_{0,2}, \tag{4.23b}$$

$$\rho_3 = v. \tag{4.23c}$$

Signals ρ_i are assumed to have smooth properties. Moreover, assumption on $\alpha_{0,i} = \alpha_i$ is considered. The resulting LPV model for control design purposes is

$$\dot{x} = A(\rho_1, \rho_2, \rho_3)x + B(\rho_1, \rho_2, \rho_3)u + W(\rho_1, \rho_2, \rho_3). \tag{4.24}$$

This model includes the disturbance W that has to be managed through a suitable control intervention. Therefore, the control intervention can be divided into two terms, such as

$$a_y = a_{y,0} + a_{y,1}, \tag{4.25}$$

where the control movement $a_{y,0}$ is intended to handle the disturbance W, while the control movement $a_{y,1}$ ensures performance (10.18). $a_{y,0}$ is a feedforward term of the control input, which is determined by the first components in B, W:

$$a_{y,0} = -\frac{W_{1,1}}{B_{1,1}} = -\frac{F_{0,1}(\rho_1)l_1 - F_{0,2}(\rho_2)l_2}{c_1(\rho_1)l_1}. \tag{4.26}$$

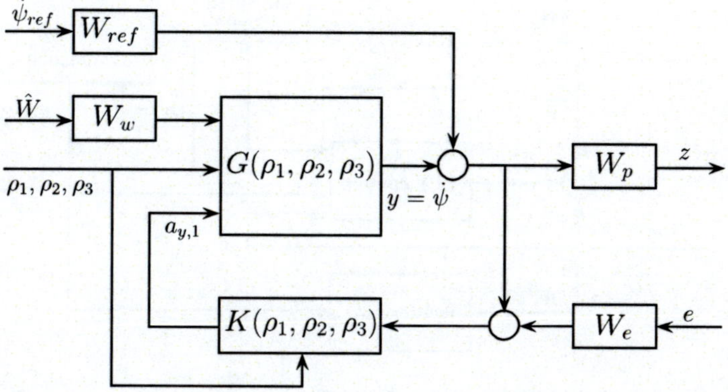

Fig. 4.9 Illustration of the closed-loop system

Applying the former equation, the static disturbance of the system W can be reduced to the following expression:

$$\hat{W}(\rho_2, \rho_3) = \begin{bmatrix} 0 \\ \frac{F_{0,2}(\rho_2)(l_1+l_2)}{ml_1\rho_3} \end{bmatrix}, \tag{4.27}$$

which depends on ρ_2 and ρ_3.

The altered control-oriented LPV model is given as

$$\dot{x} = A(\rho_1, \rho_2)x + B(\rho_1, \rho_2)a_{y,1} + \hat{W}(\rho_2, \rho_3). \tag{4.28}$$

The resulting system has the control intervention $a_{y,1}$. Through this actuation the performance requirement in (4.22) must be guaranteed, and simultaneously, to cancel $\hat{W}$. The controller uses the measured $\dot{\psi}$ for computing $a_{y,1}$. Figure 4.9 shows the closed-loop architecture of the system.

LPV control design uses more weighting functions that scale the frequency and amplitude ranges of input and output signals. Weighting function $W_{ref} = \frac{A_1}{T_1 s+1}$ scales the reference signal $\dot{\psi}_{ref}$, whose parameters A_1, T_1 are constant. The first-order form guarantees a low-pass filter capability, i.e., sharp variations in the reference signal are refined. W_w represents the scaling of the signal $\hat{W}$, which is considered to be a disturbance in this setup. Due to its low variation, a constant form $W_w = A_2$ can be applied in the synthesis process. $W_e = \frac{A_3}{T_2 s+1}$ scales the noise of the sensor on the e measurement, which also includes a rapid change in the sensor noise signal. Next, $w = \begin{bmatrix} \hat{W} & e \end{bmatrix}^T$ determines the disturbance. By the W_p weighting function, the tracking capability of the closed-loop system can be improved. It is selected in the second-order form $W_p = \frac{A_4 s+A_5}{T_4 s^2+T_3 s+1}$, in which form T_3, T_4 and A_4, A_5 are selected design parameters.

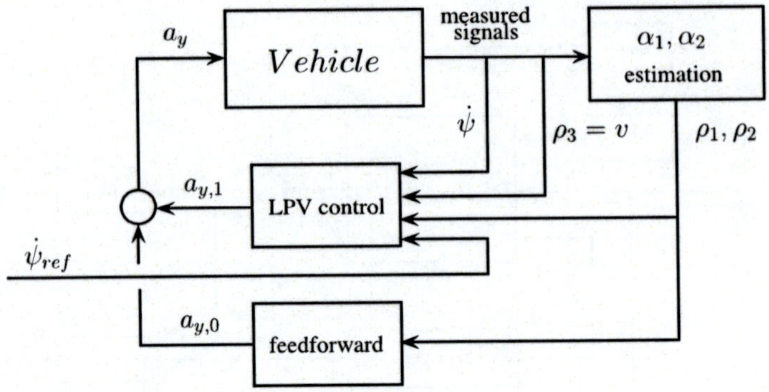

Fig. 4.10 Illustration of control system architecture

The control design is based on the LPV method using parameter-dependent Lyapunov functions, as described in Bokor and Balas (2005) and Packard and Balas (1997). The quadratic LPV performance problem is to select the parameter-varying controller $K(\rho_1, \rho_2, \rho_3)$, with which quadratic stability of the closed-loop system is achieved. Moreover, from w to z the computed induced $\mathscr{L}_2$ norm of the LPV system $G_{F_\mathscr{P}}$, with zero initial conditions, must be smaller than a predefined γ:

$$\|G_{F_\mathscr{P}}\|_\infty = \sup_{\rho \in F_\mathscr{P}} \sup_{\|w\|_2 \neq 0, w \in \mathscr{L}_2} \frac{\|z\|_2}{\|w\|_2} < \gamma. \tag{4.29}$$

The solution of the control problem leads to a problem of linear matrix inequality (LMIs), i.e., stability and the condition of (6.12) can be guaranteed if there exists an $X(\rho) > 0$ fulfilling the following LMI for all ρ:

$$\begin{bmatrix} A_{cl}^T X + X A_{cl} + d/dt(X) & X B_{cl} & \gamma^{-1} C_{cl}^T \\ B_{cl}^T X & -I & \gamma^{-1} D_{cl}^T \\ \gamma^{-1} C_{cl} & \gamma^{-1} D_{cl} & -I \end{bmatrix} < 0. \tag{4.30}$$

If there is an existing solution, $x^T X(\rho) x$ is a ρ-dependent Lyapunov function for the closed-loop at all ρ; see Yu and Sideris (1997) and Packard and Balas (1997).

The scheme of the whole control system is depicted in Fig. 4.10. For the feedback LPV-based control, measurements on ρ signals are also needed: side slip angles, and longitudinal velocity can be estimated from some measured signals (Song et al. 2002; Venhovens and Naab 1999; Baffet et al. 2006). Values of scheduling variables are also involved in the feedforward part of the control system. Moreover, the components of $\dot{\psi}_{ref}$ and the actual $\dot{\psi}$ must also be measured. The resulting overall control input of the suspension is $a_y = a_{y,0} + a_{y,1}$.

4.5 Demonstration Example

The effectiveness of the control system of variable-geometry vehicle suspension is demonstrated through different scenarios. In the examples, double-lane change maneuver with $v = 130$ km/h constant velocity has to be performed. The selected full-size vehicle has a 250 kW engine with a 7-gear transmission. The vehicle has independent suspension on the front and rear axles as well. The parameters of the vehicle are given in Table 4.1. In the simulations, the driver model is based on MacAdam (1981). The LPV design weighting functions are selected as follows:

$$W_p = \frac{0.35s + 5}{s^2 + 12s + 20}, \tag{4.31a}$$

$$W_{ref} = \frac{1}{0.01s + 1}, \tag{4.31b}$$

$$W_e = \frac{0.005}{0.001s + 1}, \tag{4.31c}$$

$$W_w = 0.01. \tag{4.31d}$$

The presented demonstrations have been realized in a software-in-the-loop (SIL) environment, as depicted in Fig. 4.11. The SIL setup contains a PC with a vehicle dynamic simulation software and dSPACE-AutoBox rapid prototyping system. In the PC standard industrial tools, i.e., CarSim® vehicle simulator and MATLAB® software are involved. The suspension control on the dSPACE equipment is realized, which is a conventional environment for vehicle control test setups. The control intervention based on the differential equations of the dynamic LPV controller through a discrete-time solver with 0.01 s sampling time is computed. The communication between dSPACE and PC through the CAN bus is implemented.

In the case of double-lane change scenarios, the motion of the vehicle usually in the last stage is critical, because the boundaries of the lane by the vehicle are reached. The effectiveness of the control can be demonstrated by keeping the vehicle inside of the lane along its entire route. Figure 4.12 depicts the motion of the two simulated vehicles in the critical last section from above. Critical points are at the 150 m and 180 m locations, depicted with red-colored ellipsoids. In the first scenario,

Table 4.1 Parameters of tire and vehicle models

m	1833 kg	$c_{1,1}$	3110.52	$c_{1,2}$	2662.605
J	2765 kgm^2	$c_{2,1}$	−12.46	$c_{2,2}$	−10.778
l_1	1.402 m	$c_{3,1}$	−30.86	$c_{3,2}$	−17.528
l_2	1.646 m	$c_{4,1}$	0.5	$c_{4,2}$	0.284
g_0	8848.317	$c_{5,1}$	0.15	$c_{5,2}$	0.085
g_4	−0.442	$c_{6,1}$	−0.005	$c_{6,2}$	−0.003

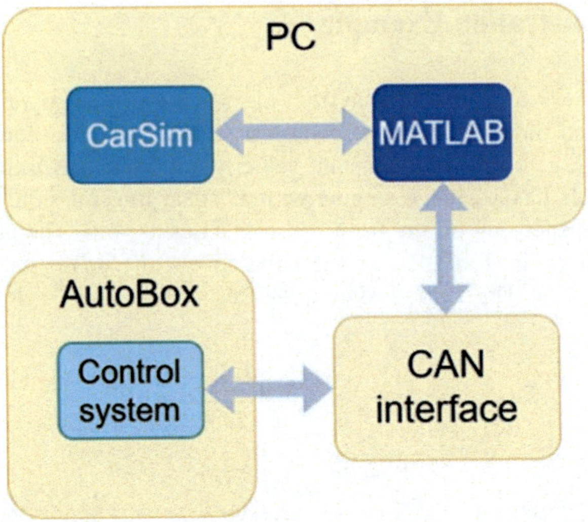

Fig. 4.11 Software-in-the-loop simulation setup

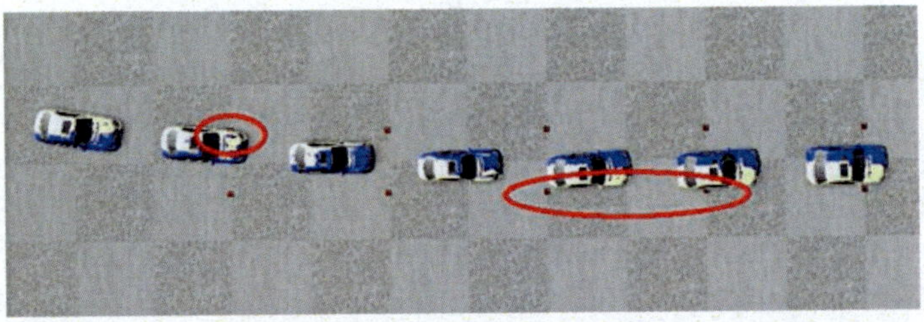

Fig. 4.12 Illustration of the path of uncontrolled and controlled vehicles

the vehicle (blue) has the designed LPV controller. The uncontrolled vehicle in the second scenario is illustrated with gold color, i.e., pure driver steering without control intervention is carried out. It can be seen that the edge line with the front left wheel of the uncontrolled vehicle is crossed; see the first ellipsoid (entry of the last route section). Furthermore, at the end of the route the uncontrolled vehicle is close to the edge line; see its rear right wheel (second ellipsoid). Nevertheless, due to the appropriate control intervention the maneuver is performed in the first scenario, and adequate vehicle dynamics by the control system is resulted; see Fig. 4.12.

Figure 4.13 illustrates the results of two additional scenarios, i.e., when the vehicle is controlled through tilting and steering or through pure tilting. It can be seen that pure steering intervention is not sufficient enough to carry it out (Fig. 4.13a, b), and consequently, the edge line of the route entrance is crossed, which is illustrated in Fig. 4.13c. Nevertheless, Fig. 4.13b shows that a vehicle with pure tilting intervention executes the cornering, but its lateral movement at the straight line has an increased overshoot and the oscillation has a long duration, as described in Fig. 4.13b. Distance

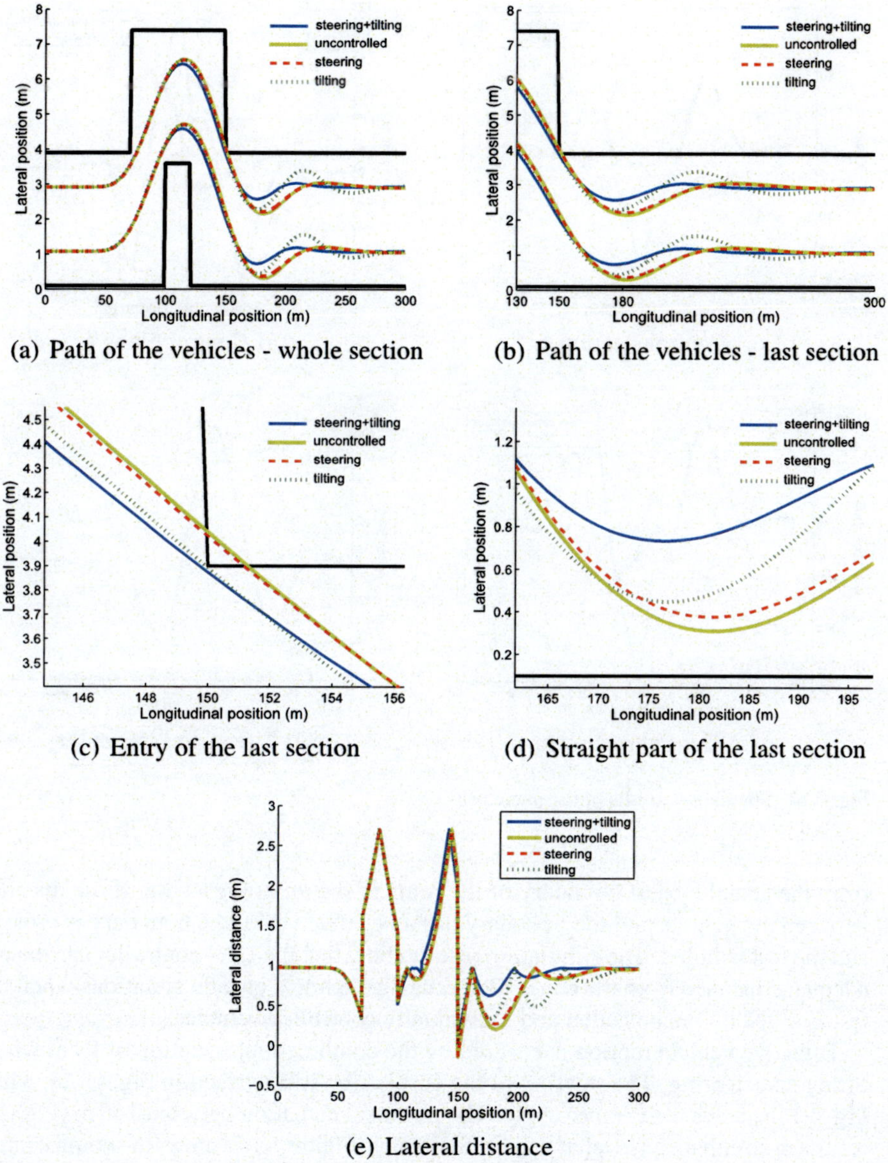

(a) Path of the vehicles - whole section

(b) Path of the vehicles - last section

(c) Entry of the last section

(d) Straight part of the last section

(e) Lateral distance

Fig. 4.13 Illustration of the vehicles' motion

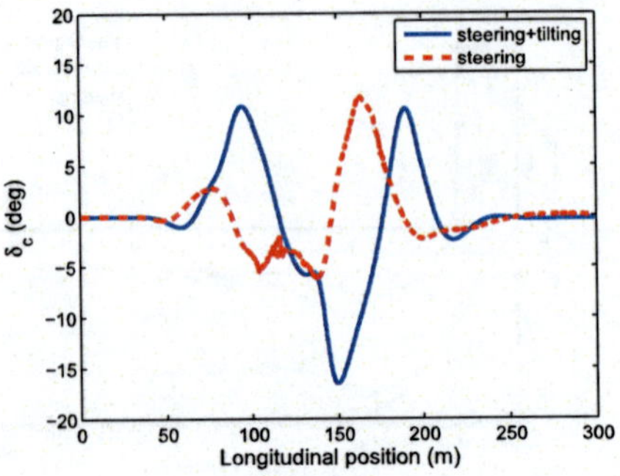

(a) Control steering δ_c

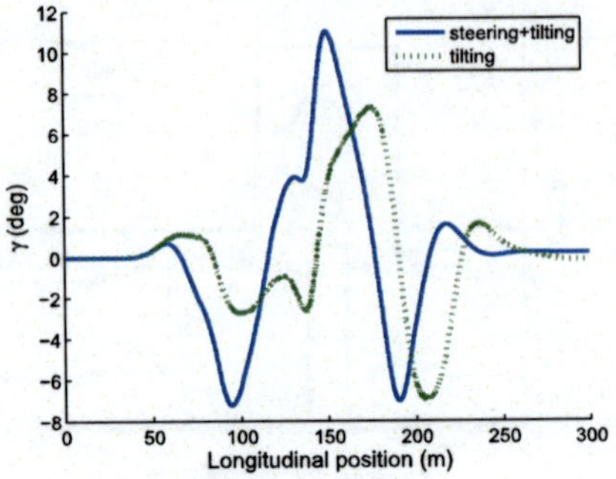

(b) Control tilting γ

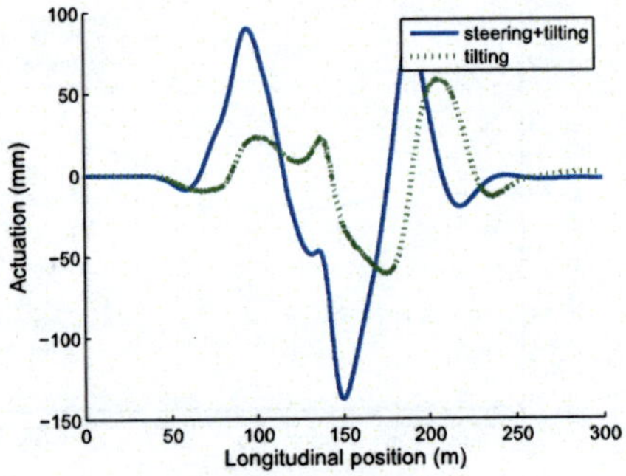

(c) Control motion a_y

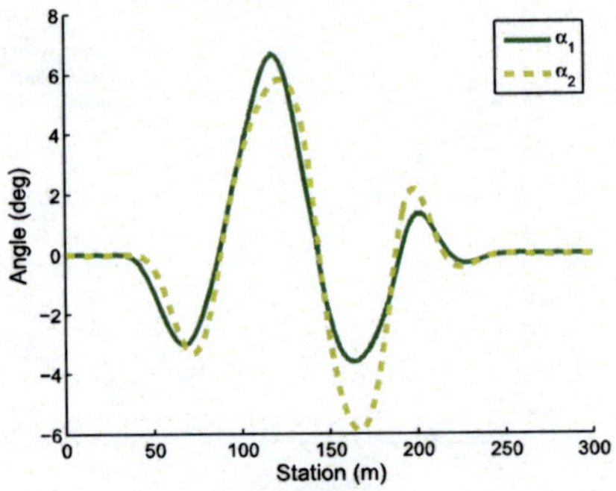

(d) Tire slip angles α_1, α_2

Fig. 4.14 Simulation results of the maneuver

from the actual closest boundary of the route is shown in Fig. 4.13e. If the distance between the vehicle and the boundary has a negative value, the boundary is crossed and the test is failed. The simulation results show that the LPV controller can ensure adequate distance from the edges along the whole track, even in situations when the test is failed by uncontrolled and individually controlled vehicles.

Effective vehicle motion is ensured by the combined intervention of front wheel tilting and steering. The steering of the suspension is described in Fig. 4.14a, while Fig. 4.14b shows the camber angle. The powerful actuation between 150 m …250 m results in an enhanced level of vehicle dynamics. Figure 4.14c shows the control intervention a_y, in which the distinguished characteristics of each actuation are found. The controller scheduling variables are shown in Fig. 4.14d. These signals indicate that the vehicle moves in the nonlinear region of the tire force characteristics, compared with Fig. 4.3. Nevertheless, the designed LPV control is able to operate effectively.

Finally, in Fig. 4.15, a vehicle cornering maneuver on icy road with 130 km/h is depicted. The route of the vehicles can be seen in Fig. 4.15a. It is demonstrated that the icy spot results in significant changes in the movement of vehicles, which leads to the departure of the lane; see the case with an uncontrolled vehicle in Fig. 4.15b.

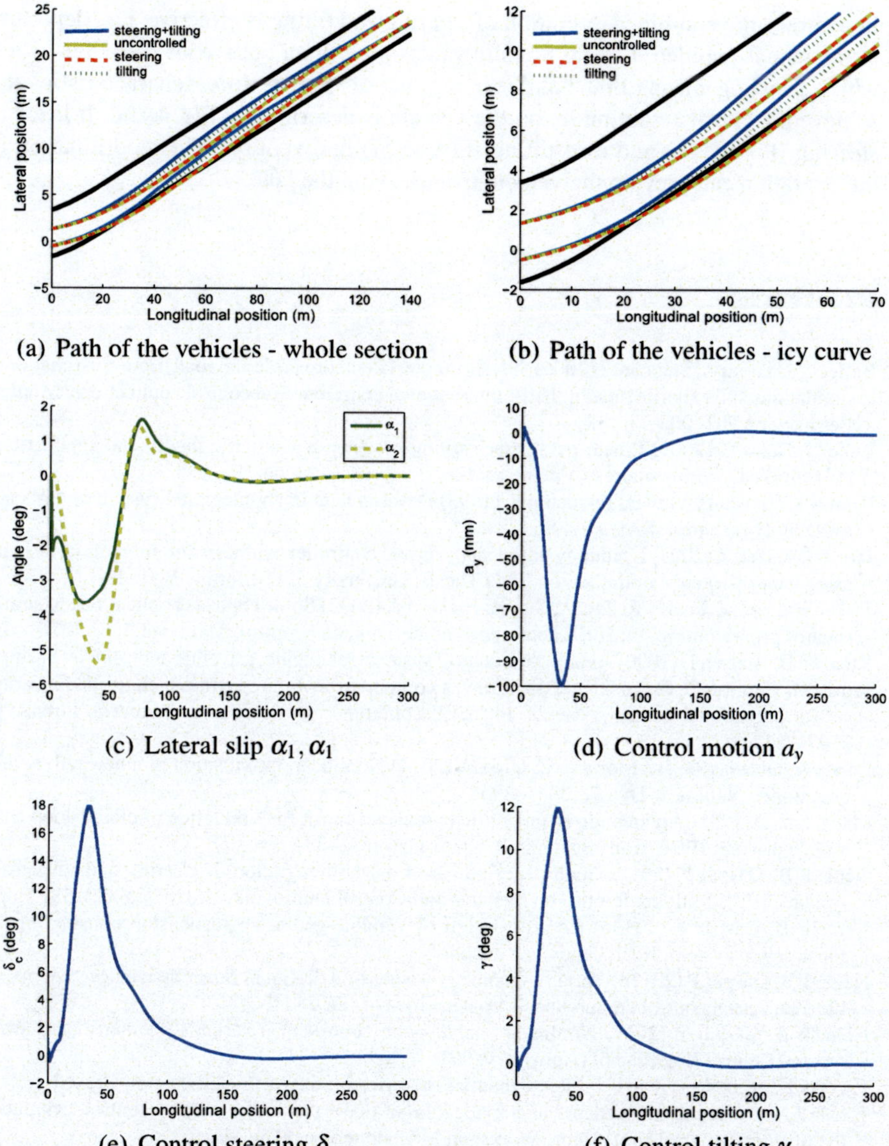

(a) Path of the vehicles - whole section

(b) Path of the vehicles - icy curve

(c) Lateral slip α_1, α_1

(d) Control motion a_y

(e) Control steering δ_c

(f) Control tilting γ

Fig. 4.15 Vehicles in the icy curve

In contrast, the combined operation of steering and tilting is effective, i.e., departure of the lane is avoided. Lateral tire slip values growth on both axles (see Fig. 4.15c), which growing means that nonlinear regions of the tire force characteristics are achieved. Control actuation a_y on the vehicle is described in Fig. 4.15d. It leads to steering (Fig. 4.15e) and also tilting (Fig. 4.15f) intervention. This coordination in the operation can prevent the vehicle from leaving the lane.

References

Baffet G, Charara A, Stéphant J (2006) Sideslip angle, lateral tire force and road friction estimation in simulations and experiments. In: IEEE international conference on control applications, Munich, Germany pp 903–908

Bokor J, Balas G (2005) Linear parameter varying systems: a geometric theory and applications. In: 16th IFAC world congress, Prague pp 1–6

Hirano Y, Harada H, Ono E, Takanami K (1993) Development of an integrated system of 4WS and 4WD by H_∞ control. SAE J 79–86

Jarvis-Wloszek Z (2003) Lyapunov based analysis and controller synthesis for polynomial systems using sum-of-squares optimization. PhD Thesis, University of California, Berkeley

Jarvis-Wloszek Z, Feeley R, Tan W, Sun K, Packard A (2003) Some controls applications of sum of squares programming. In: IEEE conference on decision and control, Maui, vol 5, pp 4676–4681

Kiencke U, Nielsen L (2000) Automotive control systems for engine, driveline and vehicle. Springer

Korda M, Henrion D, Jones CN (2013) Convex computation of the maximum controlled invariant set for polynomial control systems. In: IEEE conference on decision and control, Firenze pp 7107–7112

López A, Olazagoitia JL, Moriano C, Ortiz A (2014) Nonlinear optimization of a new polynomial tyre model. Nonlinear Dyn 78:2941–2958

MacAdam C (1981) Application of an optimal preview control for simulation of closed-loop automobile driving. IEEE Trans Syst Man Cybern 11(6):393–399

Németh B, Gáspár P (2012) Control design based on the integration of steering and suspension systems. IEEE Multiconference on Systems and Control Dubrovnik, Croatia pp 382–387

Németh B, Gáspár P (2013a) Control design of variable-geometry suspension considering the construction system. IEEE Trans Veh Technol

Németh B, Gáspár P (2013b) Variable-geometry suspension design in driver assistance systems. In: 12nd European control conference, Zürich pp 1481–1486

Németh B, Gáspár P (2017) Nonlinear analysis and control of a variable-geometry suspension system. Control Eng Prac 61(April):279–291

Pacejka HB (2004) Tyre and vehicle dynamics. Elsevier Butterworth-Heinemann, Oxford

Packard A, Balas G (1997) Theory and application of linear parameter varying control techniques. In: American control conference, Workshop I, Albuquerque, New Mexico

Parrilo P (2003) Semidefinite programming relaxations for semialgebraic problems. Math Progr B 96(2):293–320

Prajna S, Papachristodoulou A, Wu F (2004) Nonlinear control synthesis by sum of squares optimization: a Lyapunov-based approach. In: 5th IEEE Asian control conference, vol 1, pp 157–165

Rajamani R (2005) Vehicle dynamics and control. Springer

Sadri S, Wu C (2013) Stability analysis of a nonlinear vehicle model in plane motion using the concept of lyapunov exponents. Veh Syst Dyn 51(6):906–924

Scherer CW, Hol CWJ (2006) Matrix sum-of-squares relaxations for robust semi-definite programs. Math Program 107:189–211

Song C, Uchanski M, Hedrick J (2002) Vehicle speed estimation using accelerometer and wheel speed measurements. In: Proceeding of the SAE automotive transportation technology, Paris, pp 2002–01–2229

Sontag E (1989) A "universal" construction of Artstein's theorem on nonlinear stabilization. Syst & Control Lett 13:117–123

Tan W, Packard A (2008) Stability region analysis using polynomial and composite polynomial Lyapunov functions and sum-of-squares programming. IEEE Trans Autom Control 53(2):565–571

Topcu U, Packard A (2009) Local robust performance analysis for nonlinear dynamical systems. In: Proceeding of the American control conference, pp 784–789

Venhovens P, Naab K (1999) Vehicle dynamics estimation using kalman filters. Veh Syst Dyn 32:171–184

de Wit CC, Olsson H, Astrom KJ, Lischinsky P (1995) A new model for control of systems with friction. IEEE Trans Autom Control 40(3):419–425

Yu J, Sideris A (1997) $\mathscr{H}_\infty$ control with parametric Lyapunov functions. Syst Control Lett 30:57–69

Part II
Independent Steering with Variable-Geometry Suspension

Chapter 5
Modeling Variable-Geometry Suspension System

This chapter focuses on the model formulation for achieving independent steering functionality. In the model, the motion dynamics of the suspension and the orientation of the wheel are formulated, i.e., the lateral motion model of the vehicle, in which the variations of camber angle and scrub radius are considered.

5.1 Dynamical Formulation of Suspension Motion

The goal of intervention with variable-geometry vehicle suspension system is the modification of scrub radius through the setting of wheel tilting. As a consequence of the camber angle, a lateral force is generated on the tire–ground contact. As the longitudinal force rotates the wheel around its steering axis, the modification of scrub radius affects the dynamics of steering, i.e., lateral force on the wheel–road contact is generated.

In the proposed variable-geometry vehicle suspension system, the McPherson type of construction is utilized for executing wheel steering rotation. Figure 5.1 illustrates the construction of the given variable-geometry vehicle suspension. In this suspension system, the actuator among each wheel and corresponding wheel hubs is integrated. This solution has the capability to produce torque intervention M_{act} around point B in order to achieve the modification of wheel camber angle. On the other hand, on the hub the same effect is resulted by $-M_{act}$. In the considered suspension construction, the suspension itself can have rotation around the connection point A, which links the chassis and the suspension together. Furthermore, the suspension is linked with arms and joints to the chassis (see C) and to the hub (see point D), which can ensure the rotation and movement of the suspension.

Thus, there exist many forces which induce the movements of wheels and suspensions of the vehicle. From damping and compression, the induced force in the suspension (F_{susp}) can be formulated as

B. Németh and P. Gáspár, *Control of Variable-Geometry Vehicle Suspensions*, Advances in Industrial Control, https://doi.org/10.1007/978-3-031-30537-5_5

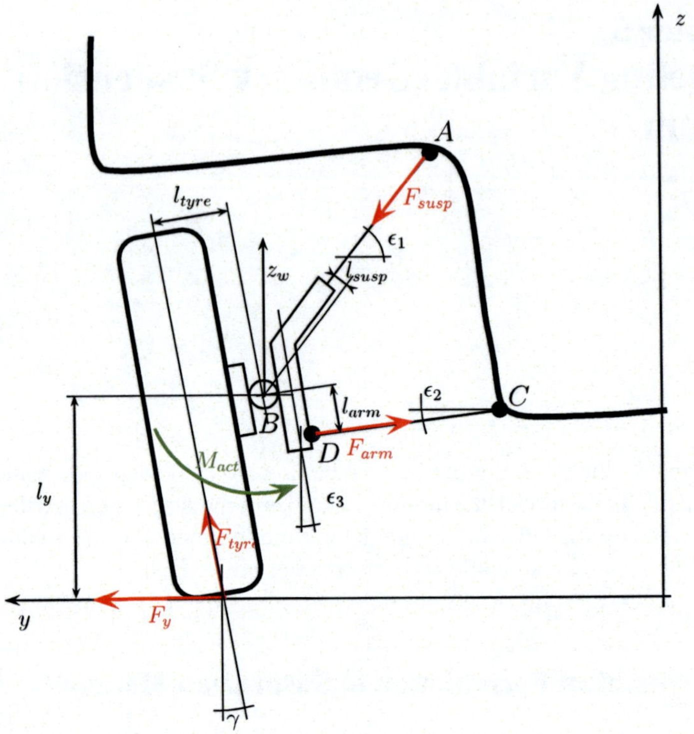

Fig. 5.1 Illustration of the construction of variable-geometry suspension (©2018 IEEE. Reprinted, with permission, from Németh et al. 2019)

$$F_{susp} = d_{susp}\frac{\dot{z}_w}{\sin\varepsilon_1} + s_{susp}\left(\frac{z_w + z_{w,0}}{\sin\varepsilon_1}\right), \tag{5.1}$$

in which expression d_{susp} is a constant parameter of damping and s_{susp} is the same for describing stiffness. Moreover, the value $z_{w,0}$ stands for a joint position, which results from the static compression of the suspension. Lateral component of the wheel–road force (F_y) can be modeled using conventional tire models; see, e.g., Pacejka (2004). F_{tyre} denotes the force component, which is generated by the compression of the tire, acting in the wheel's vertical orientation:

$$F_{tyre} = s_{tyre}\frac{z_{tyre,0} + (r_w\cos\gamma - r_w - l_{tyre}\sin\gamma - z_w)}{\cos\gamma}, \tag{5.2}$$

in which r_w represents the radius of the wheel, s_{tyre} is the stiffness coefficient of the tire and, also, tire static compression has the value $z_{tyre,0}$.

Suspension movement can be formed through the following motion equations:

$$m_{susp}\ddot{z}_w = -F_{susp} \sin \varepsilon_1 + F_{tyre} \cos \gamma + F_{arm} \sin \varepsilon_2, \tag{5.3a}$$

$$J_{susp}\ddot{\varepsilon}_3 = F_{susp}l_{susp} + F_{arm}l_{arm} - M_{act}, \tag{5.3b}$$

$$J_w\ddot{\gamma} = M_{act} - F_{tyre}l_{tyre} - F_y l_y, \tag{5.3c}$$

where (5.3a) expresses the motion dynamics of wheel hub z_w in the vertical direction. In (5.3b), active suspension torque (M_{act}) effects on wheel hub rotation ε_3. Wheel tilt dynamics γ is expressed through (5.3c), in which the lateral tire force arm is

$$l_y = r_w \cos \gamma - l_{tyre} \sin \gamma, \tag{5.4}$$

in which expressions l_{tyre}, r_w denote design parameters, similar to l_{arm}, l_{susp} and $\varepsilon_1, \varepsilon_2$ constants in (5.3). Although (5.3) provides a thorough representation of the motion dynamics, $\ddot{\varepsilon}_3 = 0$ can be neglected because of large force values F_{arm}, F_{susp}. Therefore, M_{act} has a slight impact on ε_3, which can be considered to be constant. Using this assumption, F_{arm} is calculated as follows:

$$F_{arm} = \frac{M_{act} - F_{susp}l_{susp}}{l_{arm}}. \tag{5.5}$$

The nonlinear motion model of variable-geometry suspension is confirmed via the complex mechanical system simulation software Simscape Multibody™. The suspension in this software platform using joints, arm and body elements has been constructed, and the variation in the signal of wheel tilting has been studied. A simulation example, in which the suspension model (5.3) and the operation of the SimMechanics model are compared, can be seen in Fig. 5.2. Similar M_{act} chirp signal inputs are realized in the simulations; see Fig. 5.2a. An illustration of wheel tilting comparison can be found in Fig. 5.2b. This demonstrates that in their operations, the difference between the models is limited in the high range of the operation.

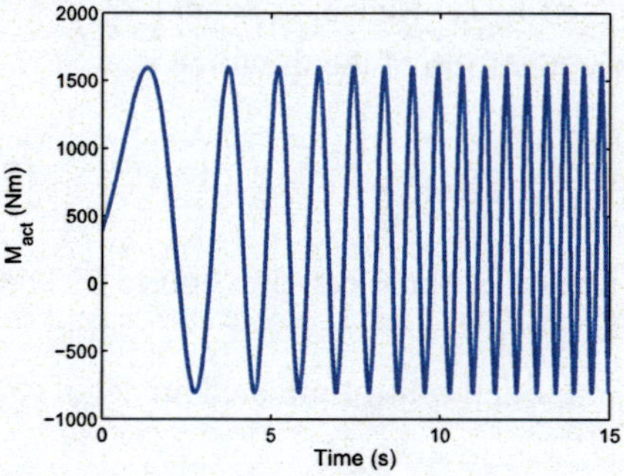

(a) Control input M_{act}

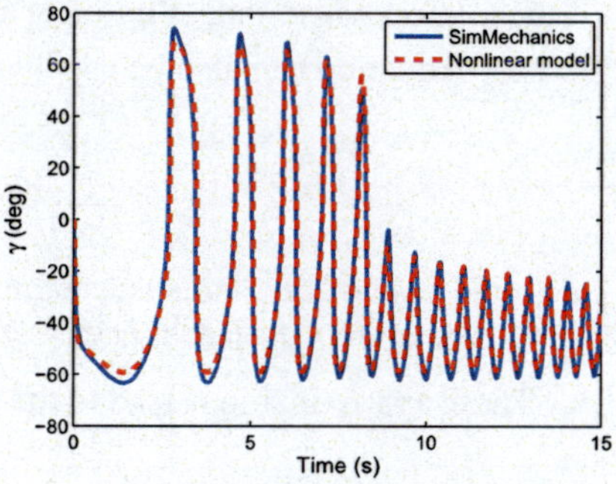

(b) Wheel tilting γ

Fig. 5.2 Validation of the model using a chirp-formed signal

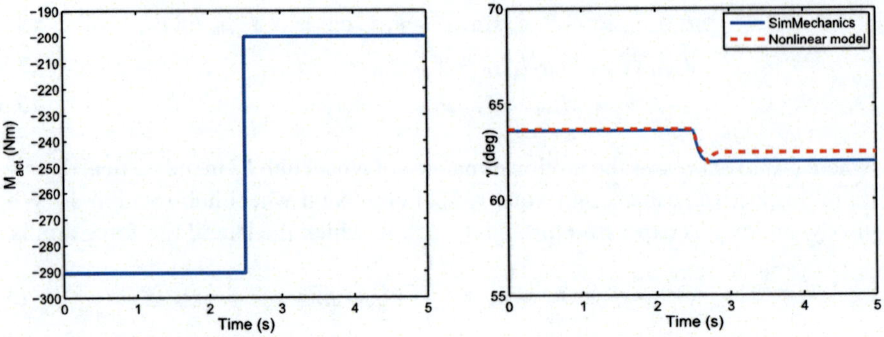

(a) Control input M_{act}

(b) Camber angle γ

Fig. 5.3 Model validation—step signal

An additional example of validating the nonlinear suspension model is illustrated in Fig. 5.3. In this case, M_{act} with a step form on the system is actuated; see Fig. 5.3a. The camber angle also changes as a result of the torque variation; see Fig. 5.3b. The results show that the operations of the models are close to each other, i.e., the error of γ in steady state is under 2%. Consequently, the proposed nonlinear model describes the motion of the suspension appropriately.

The main goal of the proposed nonlinear model is to transform it into a control-oriented model, which is capable of the design of the variable-geometry suspension controller. Thus, (5.3) nonlinear suspension model is converted to a linear state-space form. The following assumptions for achieving a successful linearization process are taken:

- During the control-oriented model formulation small tilting angles are assumed, and this consideration results in $\sin\gamma = \gamma$, $\cos\gamma = 1$ approximations for (5.3).
- Due to limited value of γ, F_y through a linearized formula can be approximated: $F_y = C\alpha$, where side-slip angle is denoted by α. Hence, $\alpha = \tan\left(\frac{r_w\dot{\gamma}}{v}\right) \approx \frac{r_w\dot{\gamma}}{v}$ approximation can also be used, and, thus, the form of the tire force is

$$F_y = C\frac{r_w\dot{\gamma}}{v}. \tag{5.6}$$

- In the case of tire and suspension, the impact of static compression in (5.1) and (5.2) can be neglected.

These approximations lead to the next simplified control-oriented model of (5.3)

$$m_{susp}\ddot{z}_w = \left(1 + \frac{l_{susp}}{l_{arm}}\frac{\sin\varepsilon_2}{\sin\varepsilon_1}\right)(-s_{susp}z_w - d_{susp}\dot{z}_w) + s_{tyre}(-l_{tyre}\gamma - z_w) + \frac{\sin\varepsilon_2}{l_{arm}}M_{act}, \tag{5.7a}$$

$$J_w\ddot{\gamma} = M_{act} - C\frac{r_w\dot{\gamma}}{v}(r_w - l_{tyre}\gamma) - s_{tyre}l_{tyre}(-\gamma l_{tyre} - z_w). \tag{5.7b}$$

Apparently, (5.7b) contains a nonlinear term $C\frac{r_w\dot{\gamma}}{v}\cdot l_{tyre}\gamma$. Another consideration is that the assumption of low γ values; this term can be eliminated. Accordingly, the state-space form is yielded as

$$\dot{x}_{susp} = A_{susp}x_{susp} + B_{susp}u_{susp}, \tag{5.8}$$

which has the vector of states $x = \left[\dot{z}_w\ z_w\ \dot{\gamma}\ \gamma\right]^T$ and $u = M_{act}$. Matrices A_{susp}, B_{susp} are constructed as

$$A_{susp} = \begin{bmatrix} -\frac{d_{susp}\eta}{m_{susp}} & -\frac{s_{tyre}+s_{susp}\eta}{m_{susp}} & 0 & -\frac{s_{tyre}l_{tyre}}{m_{susp}} \\ 1 & 0 & 0 & 0 \\ 0 & \frac{s_{tyre}l_{tyre}}{J_w} & \frac{Cr_w^2}{J_w v} & \frac{s_{tyre}l_{tyre}^2}{J_w} \\ 0 & 0 & 1 & 0 \end{bmatrix},$$

$$B_{susp} = \begin{bmatrix} \frac{\sin\varepsilon_2}{m_{susp}l_{arm}} \\ 0 \\ \frac{1}{J_w} \\ 0 \end{bmatrix},$$

where $\eta = \left(1 + \frac{l_{susp}}{l_{arm}}\frac{\sin\varepsilon_2}{\sin\varepsilon_1}\right)$.

5.2 Modeling Lateral Dynamics Considering Variable-Geometry Vehicle Suspensions

The formulation on the dynamics of variable-geometry suspension can lead to its integration into the lateral motion model of the vehicle. Thus, the previous formulation provides the possibility of a deeper insight into the dynamics of steering. The lateral dynamics is described by the subsequent equations, see Rajamani (2005):

$$J\ddot{\psi} = C_1 l_1\left(\delta - \beta - \frac{\dot{\psi}l_1}{v}\right) - C_2 l_2\left(-\beta + \frac{\dot{\psi}l_2}{v}\right), \tag{5.9}$$

$$mv(\dot{\psi} + \dot{\beta}) = C_1\left(\delta - \beta - \frac{\dot{\psi}l_1}{v}\right) + C_2\left(-\beta + \frac{\dot{\psi}l_2}{v}\right), \tag{5.10}$$

in which m denotes vehicle mass, C_1, C_2 represent stiffness of cornering, J is yaw inertia and distances l_1, l_2 are constant parameters. The time-dependent signals of the system are yaw angle (ψ) and vehicle side slip β.

Rotation dynamics of the steered wheel is expressed by

$$\ddot{\delta}_i = \frac{r_{\delta,i}F_{l,i}}{J_{\delta,i}}, \tag{5.11}$$

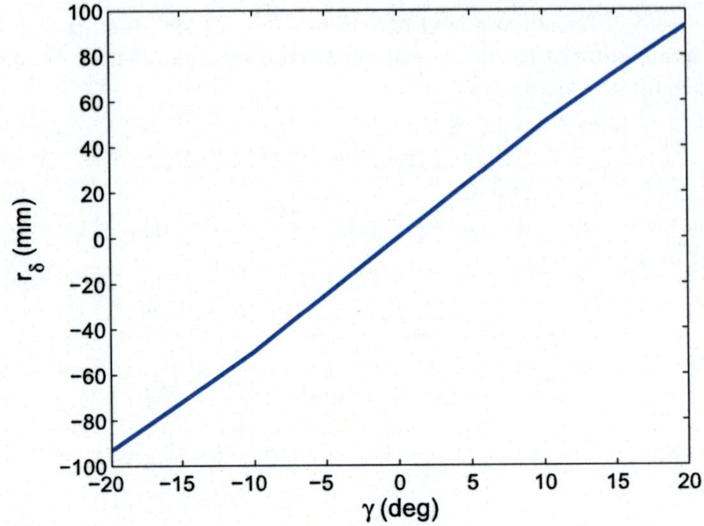

Fig. 5.4 Relationship of camber angle and scrub radius

where i stands for right $i = r$ or left $i = l$ wheel, and $J_{\delta,i}$ denotes steering inertia value for the given wheel. $F_{l,i}$ is braking/traction force on the wheel and $r_{\delta,i}$ is the scrub radius of the wheel. In addition, $r_{\delta,i}$ depends on γ_i, which through analysis on SimMechanics simulations can be approximated, see, e.g., Fig. 5.4:

$$r_{\delta,i} = \varepsilon \gamma_i, \tag{5.12}$$

where ε is a design parameter. Equation (5.11) is converted as

$$\ddot{\delta}_i = \frac{\varepsilon F_{l,i}}{J_{\delta,i}} \gamma_i. \tag{5.13}$$

Nonetheless, in the single-track model (5.9), δ characterizes the average steering angle of the front wheels, calculated as the average of the left and right steering wheel angles:

$$\delta = \frac{\delta_l + \delta_r}{2}. \tag{5.14}$$

Vehicle lateral dynamics are formed through (5.9), (5.13) and (5.14). The model state vector is $x_{lat} = \begin{bmatrix} \dot{\psi} & \beta & \dot{\delta}_l & \delta_l & \dot{\delta}_r & \delta_r \end{bmatrix}^T$, which results in the state-space representation:

$$\dot{x}_{lat} = A_{lat} x_{lat} + B_{lat} u_{lat}, \tag{5.15}$$

where the control input vector is $u_{lat} = \begin{bmatrix} \gamma_l & \gamma_r \end{bmatrix}^T$ and the matrices are

$$A_{lat} = \begin{bmatrix} -\frac{C_1 l_1^2 + C_2 l_2^2}{Jv} & \frac{-C_1 l_1 + C2 l_2}{J} & 0 & \frac{C_1 l_1}{2J} & 0 & \frac{C_1 l_1}{2J} \\ \frac{-C_1 l_1 + C_2 l_2}{mv^2} - 1 & -\frac{C_1 + C_2}{mv} & 0 & \frac{C_1}{2m} & 0 & \frac{C_1}{2m} \\ 0 & 0 & 1 & 0 & 0 & 0 \\ 0 & 0 & 0 & 0 & 0 & 0 \\ 0 & 0 & 0 & 0 & 1 & 0 \\ 0 & 0 & 0 & 0 & 0 & 0 \end{bmatrix},$$

$$B_{lat} = \begin{bmatrix} 0 & 0 & 0 & \frac{\varepsilon F_{l,l}}{J_{\delta,l}} & 0 & 0 \\ 0 & 0 & 0 & 0 & 0 & \frac{\varepsilon F_{l,r}}{J_{\delta,r}} \end{bmatrix}^T.$$

5.3 Model Formulation for Suspension Actuator

In the rest of this chapter, the dynamic model of a hydraulic actuator for variable-geometry suspension is presented, with which the camber angle of the suspension can be realized; see Fig. 5.5. Operation of the variable-geometry suspension is a critical vehicle control problem. The benefit of the electronic actuator type is the simple structure, which involves only an electrical power resource. Still, the electric motor has limited torque capability, and it has increased space requirements. But, with an

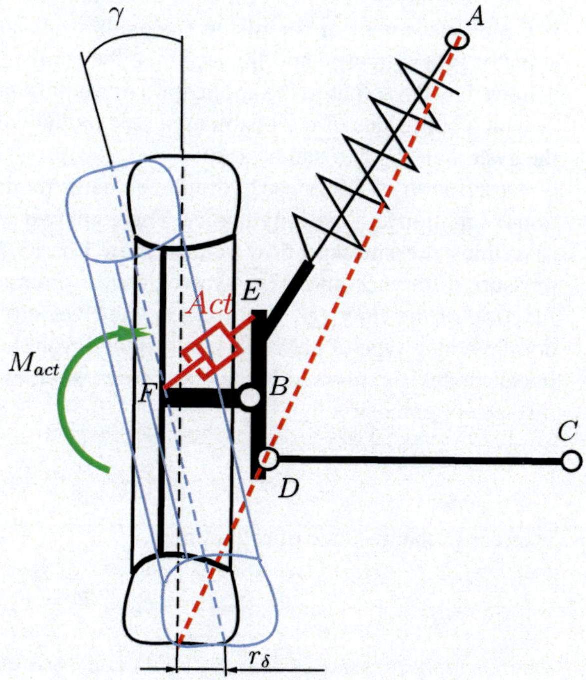

Fig. 5.5 Illustration of the suspension construction (©2017 IEEE. Reprinted, with permission, from Németh et al. 2017)

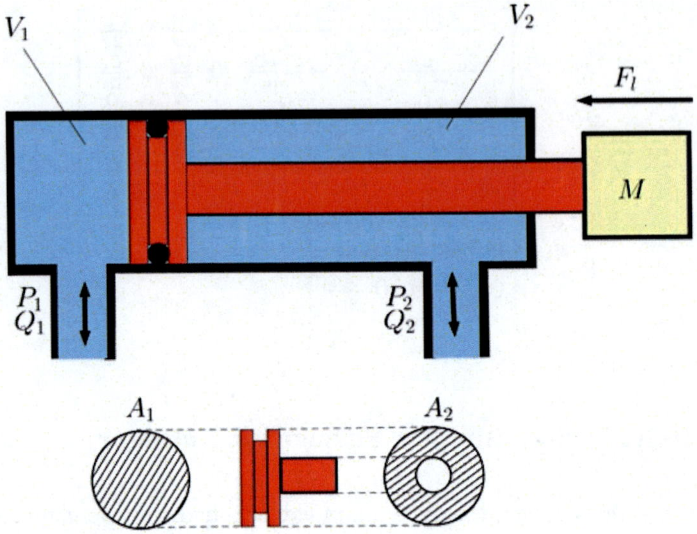

Fig. 5.6 Schematic view on hydraulic actuator construction (©2017 IEEE. Reprinted, with permission, from Németh et al. 2017)

electro-hydraulic type of actuator, increased control capability can be achieved, and the cylinder is installed in small spaces in the suspension.

The real physical control input of the actuator is valve position x_v, and its output is the displacement of the piston (Németh et al. 2015). In this study, the selected actuator is asymmetric and linear. Since the camber angle in a wide region must be achieved by the actuator, the application of a double-acting cylinder is recommended, in which both sides of the piston are acted by the working fluid. The architecture of the system in Fig. 5.6 can be seen.

Description of nonlinear hydraulic actuator model can be originated from equations of motion and hydrodynamics. There are two types of hydrodynamic equations describing pressure and flow values. Flow Eqs. (5.18b) and (5.18d) rely upon ΔP pressure difference and $A(x_v)$ valve surface. Pressure Eqs. (5.18a), (5.18c) are the function of the flow Q_i, position (x_p) and velocity ($\dot{x}_p$) of the piston. Due to the double-acting type of actuator, it is supplied from both sides. Accordingly, the nonlinear model is expressed for two instances, such as extension and retraction. In the instance of extension:

$$P_i = P_s, \quad P_l = P_0, \tag{5.16}$$

whereas in the instance of retraction:

$$P_i = P_0, \quad P_l = P_s, \tag{5.17}$$

where supply pressure is denoted by P_s and P_0 is environmental pressure.

Motion equations of the piston are formulated as (Meritt 1967; Šulc and Jan 2002)

$$\dot{P}_1 = \frac{B_f}{A_1 x_p + V_1}(-Q_1 - \alpha(P_1 - P_2) - A_1 \dot{x}_p), \tag{5.18a}$$

$$Q_1 = C_f A(x_v)\sqrt{\frac{2|P_i - P_1|}{\rho}}, \tag{5.18b}$$

$$\dot{P}_2 = \frac{B_f}{A_2(l - x_p) + V_2}(Q_2 + \alpha(P_1 - P_2) + A_2 \dot{x}_p), \tag{5.18c}$$

$$Q_2 = C_f A(x_v)\sqrt{\frac{2|P_2 - P_l|}{\rho}}, \tag{5.18d}$$

$$m\ddot{x}_p = F_l - b\dot{x}_p + A_1 P_1 - A_1 P_2, \tag{5.18e}$$

where P_1 and P_2 are chamber pressures, and B_f is the Bulk modulus. External load on the cylinder is F_l, α denotes the coefficient of internal cross-port leakage and V_i is the initial volume. Furthermore, the maximal displacement of the piston is l, and two coefficients in the model are b on the viscous damping coefficient and C_f on leakage. The density of fluid is ρ, the mass of the piston is m and its position is x_p. The connections between extending and retracting motions involve a number of nonlinear effects, which requires to taken simplifications during the formulation of a control-oriented model:

- Pressure dynamics Eqs. (5.18a), (5.18c) are linearized around

$$x_0 = x_p\big|_{\gamma=0}, \tag{5.19}$$

  where x_0 is the position of the piston at zero camber angle. Applying this simplification, the first terms in the representation have constant values:

$$\eta_1 = \frac{B_f}{A_1 x_0 + V_1}, \tag{5.20}$$

$$\eta_2 = \frac{B_f}{A_2(l - x_0) + V_2}. \tag{5.21}$$

- Linearization of flow Eqs. (5.18b), (5.18d) is yielded through Taylor series expansion, i.e., linearization around $P_{i,0}$ and A_0 is carried out:

$$Q_i \approx K_{A,i} A_i + K_{P,i} P_i, \tag{5.22}$$

  where the coefficients are

$$K_{A,i} = \left.\frac{\partial Q_i}{\partial A}\right|_{A_0, P_{i,0}} = C_f \sqrt{\frac{2\triangle P_{i,0}}{\rho}}, \tag{5.23}$$

$$K_{P,i} = \left.\frac{\partial Q_i}{\partial P_i}\right|_{A_0, P_{i,0}} = \frac{C_f A_0}{\sqrt{2\triangle P_{i,0}\rho}}. \tag{5.24}$$

Pressure difference is denoted by $\triangle P_{i,0}$, which has two distinct cases.

In the case of extending motion:

$$\triangle P_{1,0} = |P_s - P_{1,0}|, \quad \triangle P_{2,0} = |P_{2,0} - P_0|. \tag{5.25}$$

In the case of retracting:

$$\triangle P_{1,0} = |P_{1,0} - P_0|, \quad \triangle P_{2,0} = |P_s - P_{2,0}|. \tag{5.26}$$

While the values of $\triangle P_{1,0}$ and $\triangle P_{2,0}$ are different in the two instances, they are very closely matched. Accordingly, the $\triangle P_{i,0}$ can be treated as a constant. With these assumptions, the next control-oriented linear form is derived:

$$\dot{P}_1 = \eta_1 K_{P,1} P_1 - \eta_1 \alpha P_1 + \eta_1 \alpha P_2 - \eta_1 A_1 \dot{x}_p - \eta_1 K_{A,1} A(x_v), \tag{5.27}$$

$$\dot{P}_2 = \eta_2 \alpha P_1 + \eta_2 K_{P,2} P_2 - \eta_2 \alpha P_2 - \eta_2 A_2 \dot{x}_p + \eta_2 K_{A,2} A(x_v), \tag{5.28}$$

$$\ddot{x}_p = \frac{A_1}{m} P_1 - \frac{A_2}{m} P_2 - \frac{b}{m} \dot{x}_p. \tag{5.29}$$

The linearized relationships are used for the formulation of a state-space description. Vector of the states is composed as $x_{hyd} = \left[P_1 \; P_2 \; \dot{x}_p \; x_p\right]^T$, while the control input is $u_{hyd} = A(x_v)$. The state equation is expressed as follows:

$$\dot{x}_{hyd} = A_{hyd} x_{hyd} + B_{hyd} u_{hyd}, \tag{5.30}$$

where

$$A_{hyd} = \begin{bmatrix} \eta_1(K_{P,1} - \alpha) & \eta_1 \alpha & -\eta_1 A_1 & 0 \\ \eta_2 \alpha & \eta_2(K_{P,2} - \alpha) & -\eta_2 A_2 & 0 \\ \frac{A_1}{m} & -\frac{A_2}{m} & -\frac{b}{m} & 0 \\ 0 & 0 & 1 & 0 \end{bmatrix},$$

$$B_{hyd} = \begin{bmatrix} -\eta_1 K_{A,1} \\ \eta_2 K_{A,2} \\ 0 \\ 0 \end{bmatrix}.$$

As can be seen, there is a strong relationship between control input and valve position. This results in $u_{hyd} = 0$ standing for the closed position of the valve, i.e., the flow of fluid is prohibited, while $u_{hyd} \neq 0$ represents an open valve. The resulting linear formula characterizes electro-hydraulic actuator dynamics, including valve position and cylinder dynamics.

References

Meritt HE (1967) Hydraulic control systems. Wiley, New York

Németh B, Varga B, Gáspár P (2015) Hierarchical design of an electro-hydraulic actuator based on robust LPV methods. Int J Control 88(8):1429–1440

Németh B, Fényes D, Gáspár P, Bokor J (2017) Control design of an electro-hydraulic actuator for variable-geometry suspension systems. In: 2017 25th Mediterranean conference on control and automation (MED), pp 180–185

Németh B, Fényes D, Gáspár P, Bokor J (2019) Coordination of independent steering and torque vectoring in a variable-geometry suspension system. IEEE Trans Control Syst Technol 27(5):2209–2220

Pacejka HB (2004) Tyre and vehicle dynamics. Elsevier Butterworth-Heinemann, Oxford

Rajamani R (2005) Vehicle dynamics and control. Springer

Šulc B, Jan JA (2002) Non linear modelling and control of hydraulic actuators. Acta Polytechnica 42(3):41–47

Chapter 6
Hierarchical Control Design Method for Vehicle Suspensions

It has been revealed from the previous section that numerous subsystems are required for providing an accurate model formulation of variable-geometry vehicle suspensions. Since the dynamics of these systems can be different, e.g., time response, delay, fastness and mathematical structure, it can be beneficial to join them in a hierarchical framework; see Sename et al. (2013) and Németh et al. (2015). Accordingly, the control functionalites of variable-geometry suspension are designed independently, i.e., suspension control on each wheel, their control on steering and on lateral dynamics. As Fig. 6.1 shows, the control systems are linked through control signals and references. In the suggested scheme, the steering and suspension controls on the left and right sides have similar construction. Furthermore, stability and performance must be ensured in the hierarchical design method. This chapter suggests a design structure for independent steering and torque vectoring, within the theory of robust control. Next, that lateral control for steering and suspension dynamics and the formulation of system uncertainty are presented. The control design of the in-wheel motors is given in Wu et al. (2013) and Hsiao (2015). Furthermore, the force/steering allocation and reconfiguration strategy is introduced.

6.1 Suspension Control Design for Wheel Tilting

The purpose of suspension control design is to set the required camber angle to a reference value $\gamma_{i,ref}$, i.e., wheel tilting, which has been required by the steering control on a higher level. The suspension control has to ensure accurate $\gamma_{i,ref}$ tracking, i.e., with low error and delay. The input of the system is a torque on the wheel $M_{act,i}$, which ensures the tilting of the wheel. Due to the goal of the control system, the most important performance requirement is the minimization of the camber angle error on both, i.e., left and right sides:

$$z_{susp} = \gamma_{i,ref} - \gamma_i, \quad |z_{susp}| \rightarrow min. \tag{6.1}$$

B. Németh and P. Gáspár, *Control of Variable-Geometry Vehicle Suspensions*, Advances in Industrial Control, https://doi.org/10.1007/978-3-031-30537-5_6

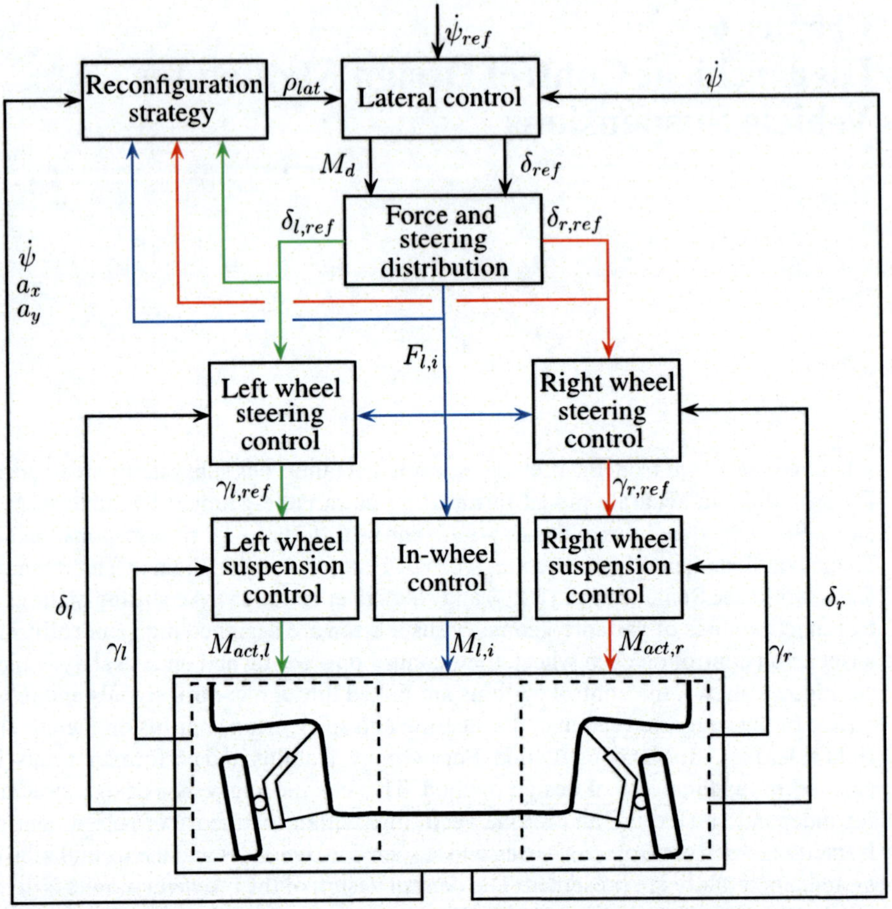

Fig. 6.1 Control system architecture (©2018 IEEE. Reprinted, with permission, from Németh et al. 2019)

The model of the system is composed of the state-space representation of the suspension (5.8) together with the predefined performance requirement, and, thus, the plant of the system is formed as

$$\dot{x}_{susp} = A_{susp}x_{susp} + B_{susp}u_{susp}, \tag{6.2a}$$

$$z_{susp} = C_{susp,1}x_{susp} + D_{susp,1}u_{susp}, \tag{6.2b}$$

$$y_{susp} = C_{susp,2}x_{susp} + D_{susp,2}w_{susp}. \tag{6.2c}$$

In the representation, it is considered that the camber angle on each wheel on the front axle can be measured. Furthermore, a sensor noise on the measurement w_{susp} is modeled in the augmented plant, which must be rejected by the suspension control. It motivates the application of $\mathscr{H}_\infty$ control design.

In the design process, it is necessary to construct controller $K_{susp,i}$ that ensures asymptotic stability of the closed-loop system and, furthermore, the transfer function from w_{susp} to z_{susp} of the closed-loop meet:

$$\left\| T_{z_{susp},w_{susp}} \right\|_\infty < \Gamma_{susp}, \tag{6.3}$$

for a predefined real positive value Γ_{susp}. The achieved control signals as a result of the design are $M_{act,l}$ and $M_{act,r}$.

6.2 Design Methods of Steering Control and Uncertainty

Steering control has the responsibility to ensure the tracking of the reference steering signal $\delta_{i,ref}$. It forms a tracking-based performance requirement against the steering control, such as

$$z_{st,i,1} = \delta_{ref} - \delta_i, \quad |z_{st,i,1}| \rightarrow min. \tag{6.4}$$

Minimization of $|z_{st,1}|$ is requested to achieve minimal control intervention:

$$z_{st,i,2} = \gamma_{i,ref}, \qquad |z_{st,i,2}| \rightarrow min. \tag{6.5}$$

The defined requirements on performances in $z_{st,i} = \left[z_{st,i,1} \; z_{st,i,2} \right]^T$ vector are formed. Thus, the control-oriented model of the system, together with equations on measurement and performance requirements, is

$$\dot{x}_{st,i} = A_{st} x_{st,i} + B_{st,i}(\rho_{st,i}) u_{st,i}, \tag{6.6a}$$
$$z_{st,i} = C_{st,1} x_{st,i} + D_{st} u_{st,i}, \tag{6.6b}$$
$$y_{st,i} = C_{st,2} x_{st,i}, \tag{6.6c}$$

where the measured signal is $y_{st} = \delta_i$. The matrix $B_{st,i}(\rho_{st,i})$ is a function of the scheduling variable $\rho_{st,i} = F_{l,i}$, which is calculated through the method of force allocation, as depicted in Fig. 6.1. Accordingly, the controllers must be designed for both the right and left wheels. Because the matrix $B_{st,i}(\rho_{st,i})$ relies upon $\rho_{st,i}$, the parameter dependence requests the use of the LPV-based design of the controller. In this method, weighting functions are used to find a balance between performances, and also to scale input signals; see the scheme of the closed-loop interconnection in Fig. 6.2.

The selection of weighting functions on input and output signals is determined by specifications on disturbance rejection and performances. The aim of $W_{st,1}$, $W_{st,2}$ weights is to ensure a trade-off between $z_{st,i,1}$, $z_{st,i,2}$ performances. Weight is selected

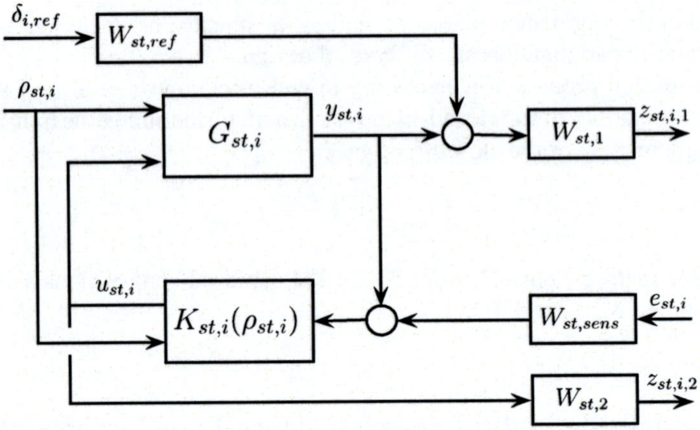

Fig. 6.2 Illustration of the structure of closed-loop interconnection (©2018 IEEE. Reprinted, with permission, from Németh et al. 2019)

as follows: $W_{st,1} = \frac{A_{st,1}}{T_{st,1}s+1}$, which influences the path following capability. Intervention values γ_i are calibrated by the weight $W_{st,2} = A_{st,2}$, with which its magnitude is defined. The role of $W_{st,ref} = A_{st,3}$ is to set the maximum value of reference signal $\delta_{i,ref}$. Moreover, $W_{st,sens} = \frac{A_{st,4}}{T_{st,4}s+1}$ includes weighting of the measurement noise on the sensor, i.e., it scales frequency range and magnitude of the measurement on steering angle error. Thus, disturbance vector involves $\delta_{i,ref}$ and $e_{st,i}$, such as $w_{st,i} = \left[e_{st,i}\ \delta_{i,ref}\right]^T$.

The control design is based on the LPV procedure, which applies parameter-dependent Lyapunov functions as detailed in Bokor and Balas (2005) and Wu et al. (1996). The quadratic LPV performance problem is as follows. It is necessary to find $K_{st,i}(\rho_{st,i})$ parameter-varying controller, which guarantees quadratic stability on the closed-loop. Moreover, the induced $\mathscr{L}_2$ norm from the disturbance to the performances must be under Γ_{st}, which is a predetermined constant with a small value. The optimization problem is formed as

$$\inf_{K_{st,i}} \sup_{\rho_{st,i}\in\mathscr{F}_{\mathscr{P}}} \sup_{\|w_{st,i}\|_2\neq 0, w_{st,i}\in\mathscr{L}_2} \frac{\|z_{st,i}\|_2}{\|w_{st,i}\|_2}. \tag{6.7}$$

The presented performance-based control design task can be depicted as a feasibility problem on a set of Linear Matrix Inequalities (LMIs); see details on its solution in Packard and Balas (1997) and Wu et al. (1996).

The role of the designed controller $K_{st,i}(\rho_{st,i})$ is illustrated in Fig. 6.1. Although it has a strong connection to steering distribution through $\delta_{i,ref}$, it is also connected to the suspension control. Thus, the overall performance level of the vehicle control is determined by all of the systems in the hierarchy, i.e., their preciseness on reference

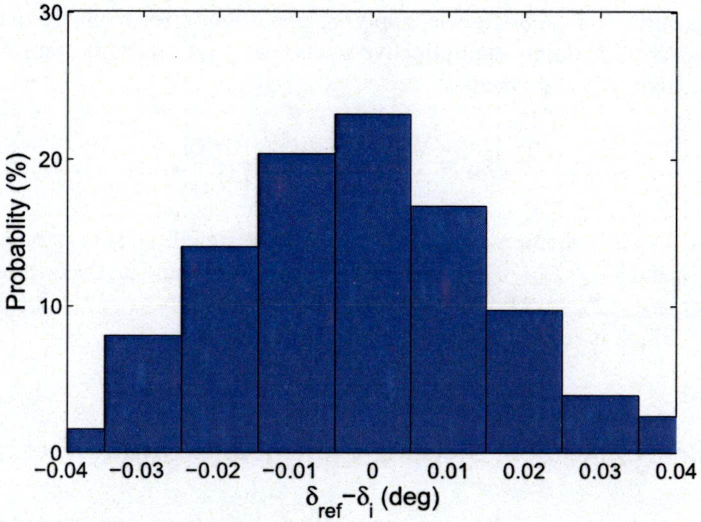

Fig. 6.3 Tracking accuracy of reference steering

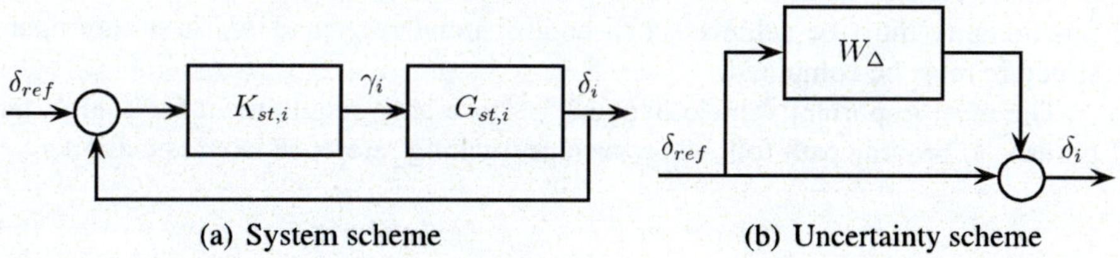

Fig. 6.4 Formulation of the steering uncertainty

signal tracking. An analysis method is proposed below, with which the tracking error on the lower-level controllers as an uncertainty can be examined.

In the method, several simulations are executed by using various initial values of $x_{st,i}(0)$, δ_{ref}. The range for each initial value is selected as $x_{st,i}(0) = \pm x_{st,max}$ and $\delta_{ref} = \pm\delta_{max}$. The intervals of $x_{st,i}$ and δ_{ref} are gridded with ε_{st} and ε_{ref} sampling. Consequently, $\left(2\frac{x_{st,max}}{\varepsilon_{st}}+1\right)^2 \cdot \left(2\frac{\delta_{max}}{\varepsilon_{ref}}+1\right)$ number of scenarios are executed. In these predefined scenarios, the low-level controller has to ensure the setting of δ_{ref} from the $x_{st,i}(0)$. In the case of each instance, the maximum value of the tracking error during the entire scenario is logged. Figure 6.3 illustrates the statistics of the examination on the maximum values, i.e., the maximum error is below 0.04°.

The achieved statistical analysis results are applied modeling uncertainties, i.e., from the viewpoint of robust control design, a worst-case scenario with the maximum of the achievable tracking error is considered. The scheme of the multiplicative uncertainty structure in Fig. 6.4 is illustrated. $G_{st,i}$ has its own dynamics, and the control $K_{st,i}$ results in error between δ_{ref} and δ_i, as depicted in Fig. 6.4a. In Fig. 6.4b, the connection of δ_{ref}, δ_i signals, applying maximum error from the examination, is expressed by forming multiplicative uncertainty, i.e., through transfer function $1 + W_\Delta$. Here, W_Δ is formed as

$$W_\Delta = \frac{\alpha_{\Delta,2}s^2 + \alpha_{\Delta,1}s + \alpha_{\Delta,0}}{T_{\Delta,2}s^2 + T_{\Delta,1}s + T_{\Delta,0}}, \tag{6.8}$$

which is for the uncertainty on the control input signal. Design parameters $\alpha_{\Delta,2}$, $\alpha_{\Delta,1}$, $\alpha_{\Delta,0}$ and $T_{\Delta,2}$, $T_{\Delta,1}$, $T_{\Delta,0}$ have the role to form a bound on the uncertainty, and, moreover, $\alpha_{\Delta,2}/T_{\Delta,2}$ is chosen consistent, Fig. 6.3. Finally, $\alpha_{\Delta,0}/T_{\Delta,0}$ is selected to be small, which is related to steady-state cases.

6.3 Coordination of Steering Control and Torque Vectoring

The goal of vehicle control is to guarantee performance requirements on the vehicle level. In the case of advanced electric vehicles, independent steering with variable-geometry vehicle suspension system and independent driving with torque vectoring functionality must be achieved, i.e., control inputs δ_{ref} and M_d in a coordinated structure must be computed.

The most important vehicle dynamic performance requirement for control purposes is to provide path following, such as tracking reference lateral position y_{ref}:

$$z_{lat,1} = y_{ref} - y, \quad |z_{lat,1}| \rightarrow min, \tag{6.9}$$

where the reference signal is determined by the road geometry. In addition, minimization of $z_{lat,1}$ must be achieved with limited control intervention $u_{lat} = \left[\delta_{ref} \; M_d\right]^T$. Accordingly, the following objectives on the actuation are formulated:

$$z_{lat,2} = \left[\delta_{ref}\right]^T, \quad |z_{lat,2}| \rightarrow min, \tag{6.10a}$$

$$z_{lat,3} = \left[M_d\right]^T, \quad |z_{lat,3}| \rightarrow min. \tag{6.10b}$$

The defined objectives, i.e., performance requirements are compressed in $z_{lat} = \left[z_{lat,1} \; z_{lat,2} \; z_{lat,3}\right]^T$ vector. Similar to the design on the lower control levels, the state-space formulation involves the dynamics of the system itself and the presented performance equations together with the measurements, such as

$$\dot{x}_{lat} = A_{lat}x_{lat} + B_{lat}u_{lat}, \tag{6.11a}$$

$$z_{lat} = C_{lat,1}x_{lat} + D_{11,lat}r_{lat} + D_{12,lat}u_{lat}, \tag{6.11b}$$

$$y_{lat} = C_{lat,2}x_{lat} + D_{21,lat}r_{lat}, \tag{6.11c}$$

where $r_{lat} = y_{ref} - y$ is noted.

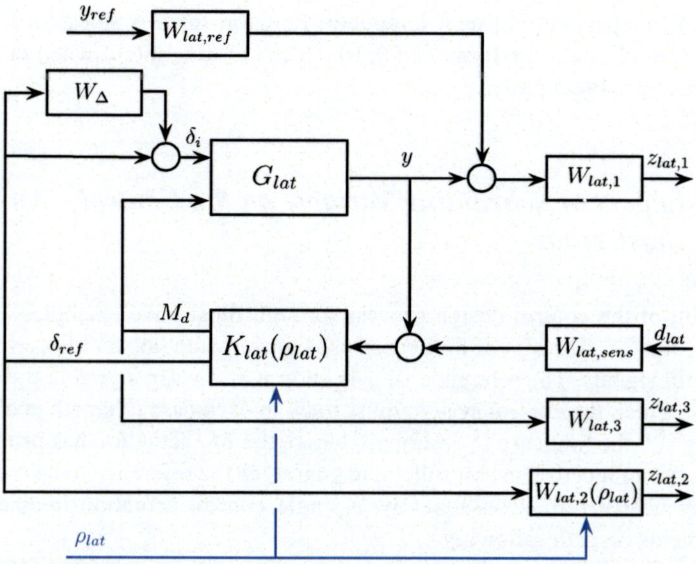

Fig. 6.5 Closed-loop interconnection structure for the lateral control design (©2018 IEEE. Reprinted, with permission, from Németh et al. 2019)

Since the formed augmented system (6.11) is linear, the coordination, i.e., balance among the control inputs δ_{ref}, M_d is achieved through a newly defined scheduling variable ρ_{lat}; see the illustration of the plant in Fig. 6.5. Three performance weighting functions are applied in the architecture. Function $W_{lat,1}$ weights the permissible error on path following, and weighting functions $W_{lat,2}(\rho_{lat})$, $W_{lat,3}$ influence the limits on the control interventions. Weighting function on steering angle is chosen as $W_{lat,2}(\rho_{lat}) = \frac{\rho_{lat}}{A_{lat,2}}$, in which expression the range of the scheduling variable is $\rho_{lat} \in [\rho_{lat,min}, \rho_{lat,max}]$. ρ_{lat} has the role to affect the intervention of the variable-geometry suspension throughout steering interference. For example, in the case of $\rho_{lat} = \rho_{lat,min}$ selection the value of $W_{lat,2}(\rho_{lat,max})$ reduces, which leads to the growth of δ_{ref}. Likewise, for $\rho_{lat} = \rho_{lat,max}$ selection the steering actuation is reduced through the high value of weight $W_{lat,2}(\rho_{lat,max})$. Despite varying weight on steering intervention, constant weight for M_d is chosen, such as $W_{lat,3} = A_{lat,3}$. It can be seen in Fig. 6.5 that W_Δ in the design is built within the formulation of (6.8). In addition, $W_{lat,ref}$ weights y_{ref} and $W_{lat,sens}$ scales noise on lateral position measurement.

The LPV-based control design problem is formed as

$$\inf_{K_{lat}} \sup_{\rho_{lat} \in \mathscr{F}_{\mathscr{P}}} \sup_{\|w_{lat}\|_2 \neq 0, w_{lat} \in \mathscr{L}_2} \frac{\|z_{lat}\|_2}{\|w_{lat}\|_2}, \tag{6.12}$$

where the disturbance vector w_{lat} contains the reference signal, the noise of the sensor and the effect of the multiplicative uncertainty. Through the optimal control design limit on ρ_{lat} is considered, and, moreover, to achieve quadratic stability a parameter-independent constant Lyapunov function form is selected. It can guarantee the stability and performance (6.9), (6.10) of the system when changing the scheduling variable rapidly.

6.3.1 Impact of Scheduling Variable on the Control—An Illustration

The result of the control design process through illustrative examples is shown in Figs. 6.6 and 6.7. The aim of the scenarios is to show the impact of ρ_{lat} selection on the control signals. The reference $\dot{\psi}_{ref}$ is chosen as a chirp signal in the simulation example, which the control system must track in each case (Németh et al. 2016a).

In Fig. 6.6 the first case is described, where the M_d actuation has primacy, while δ_i is less pronounced. The controller can guarantee the reference to be tracked while δ_l, δ_r are nearly zero. Consequently, a single control actuation is able to ensure requirements on path following.

In the case of the second example (see Fig. 6.7), torque-vectoring intervention is highly limited due to the priority on variable-geometry suspension control with $\rho_{lat} = 0.05$. The results depict the accuracy of tracking, i.e., through δ_l, δ_r an appropriate performance level can be achieved. Since both types of interventions are able to provide accurate tracking, it is possible to create a reconfiguration strategy through the appropriate selection of scheduling variables: $\rho_{lat,1}$, $\rho_{lat,2}$ and $\rho_{lat,3}$.

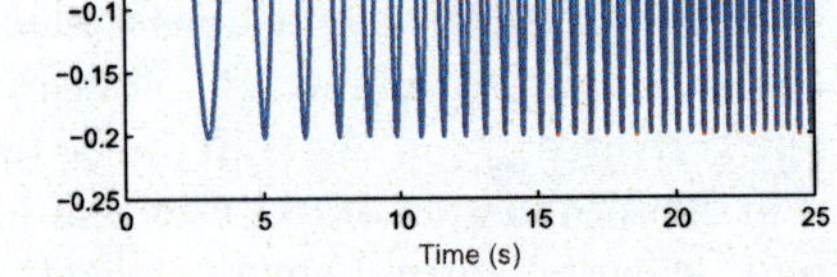

(a) Tracking of $\dot{\psi}_{ref}$

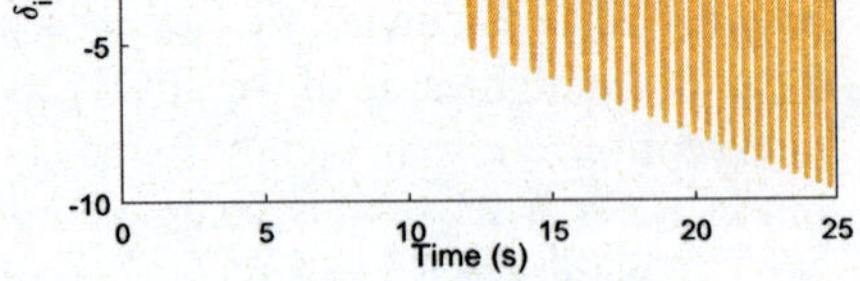

(b) Control intervention

Fig. 6.6 Simulation with selection $\rho_{lat} = 0.95$

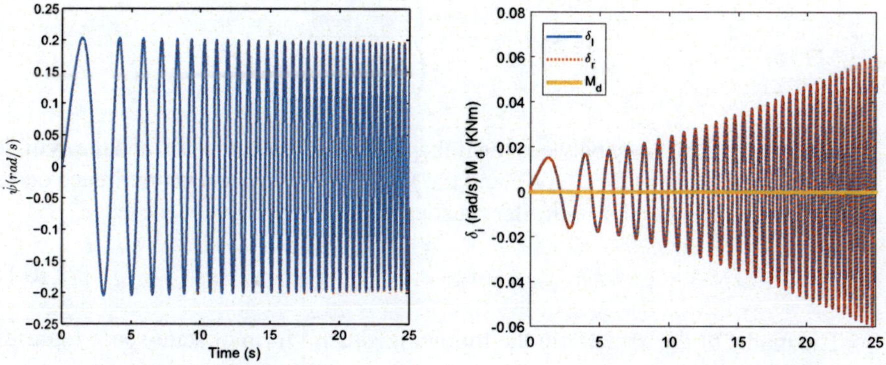

(a) Tracking of ψ_{ref} (b) Control intervention

Fig. 6.7 Simulation with selection $\rho_{lat} = 0.05$

6.4 Designing Control for Electro-hydraulic Suspension Actuator

The rest of the chapter introduces a control design method for the electro-hydraulic actuator of variable-geometry suspension system. Unfortunately, due to the special challenges of actuator control, LPV-based methods cannot be used for control purposes.

- Dynamics of the pressure can be much faster than piston pressure motion dynamics, i.e., the poles of A_{hyd} change over a large domain. Therefore, it is not possible to design actuator control and some parts of the suspension control in a joint design process.
- Control input u_{hyd} has discrete values $[A(x_{v,min}); 0; A(x_{v,max})]$, which results from the valve construction. This constraint poses further challenges for providing stability or valve positioning performance.
- Furthermore, measuring chamber pressures can be complicated or expensive in industrial applications. An effective estimation method for this purpose can be found in Németh et al. (2017).

6.4.1 The Control Design Step

Control on the valve of the system must ensure the requested camber angle by positioning the hydraulic cylinder. Because of the mechanical connection of the wheel and cylinder, the reference position of the cylinder is calculated through the geometry of the suspension:

$$x_{ref} = \frac{Lsin(\frac{\pi}{2} - \gamma_{ref})}{sin\left(\frac{\frac{\pi}{2}+\gamma_{ref}}{2}\right)}, \tag{6.13}$$

in which L is the distance between the linkage point of the lower arm and the cylinder on the hub (Németh et al. 2016b), and the value of γ_{ref} represents reference on the camber angle. Hence, the controller must ensure the reduction of the error:

$$z = x_{ref} - x_p \qquad |z| \rightarrow min. \tag{6.14}$$

The model of the actuator in the following general form of state-space representation is described as

$$\dot{x} = f(x) + gu, \tag{6.15}$$

in which vector $f(x)$ is a smooth polynomial function with the initial value $f(0) = 0$. For achieving global asymptotic stability, it is necessary to find a Control Lyapunov Function, which can be defined through the next condition.

A smooth, suitable and positive-definite function $V : \mathbb{R}^n \rightarrow \mathbb{R}$ *is a Control Lyapunov Function for system* (6.15) *if*

$$\inf_{u\in\mathbb{R}} \left\{ \frac{\partial V}{\partial x} f(x) + \frac{\partial V}{\partial x} g \cdot u \right\} < 0 \tag{6.16}$$

for each $x \neq 0$.

Finding of a Control Lyapunov Function for system (6.15) the stability can be guaranteed (Sontag 1989). Next, the tracking problem (6.14) is handled as providing stability for the system around its reference; see Németh et al. (2015). Hence, new states are determined taking into account the reference signal (6.13). The tracking error is given as

$$e = x_{ref} - x_p. \tag{6.17}$$

Assuming that the dynamics of x_p is faster than that of x_{ref}, the function of $\dot{e}$ can be simplified to

$$\dot{e} = -\dot{x}_p. \tag{6.18}$$

Applying the determined tracking error (6.17), its derivative (6.18) and the components of x_{hyd}, the following state vector of ξ can be formed:

$$\xi = \left[P_1 \; P_2 \; \alpha e \; \dot{e}\right] = \left[P_1 \; P_2 \; \alpha(x_{ref} - x_p) \; -\dot{x}_p\right]^T \tag{6.19}$$

and the derivative

$$\dot{\xi} = \left[\dot{P}_1 \; \dot{P}_2 \; \alpha\dot{e} \; \ddot{e}\right] = \left[\dot{P}_1 \; \dot{P}_2 \; -\alpha\dot{x}_p \; -\ddot{x}_p\right]^T, \tag{6.20}$$

in which expression α is a design parameter, regarding setting the rapidness of the control system. Using ξ vector, the candidate Control Lyapunov Function can be formed:

$$V = \frac{1}{2}\xi^T\xi. \tag{6.21}$$

By the derivation of (6.21), the following function holds:

$$\dot{V} = \xi^T\dot{\xi} = \left[P_1 \; P_2 \; \alpha(x_{ref} - x_p) \; -\dot{x}_p\right]\left[\dot{P}_1 \; \dot{P}_2 \; -\alpha\dot{x}_p \; -\ddot{x}_p\right]^T. \tag{6.22}$$

Apparently, in (6.22) the representations of pressure dynamics and motion dynamics on the cylinder are combined. Nevertheless, there is a significant difference in the rapidness of chamber pressure and piston dynamics, which may cause problems in the numerical computation of the controller. Accordingly, the effect of pressure dynamics on stability is not as significant as piston dynamics. Thus, the P_1, P_2 values are removed from (6.21), (6.22). It leads to the following scenarios:

- if $u_{hyd} = A(x_{v,max})$ then $P_1 = P_0$ and $P_2 = P_s$,
- if $u_{hyd} = A(x_{v,min})$ then $P_1 = P_s$ and $P_2 = P_0$ and
- if $u_{hyd} = 0$ then P_1 and P_2 have not been affected through actuation.

Consequently, Eq. (6.22) can be approximated as

$$\dot{V} \approx (-\ddot{x}_p - \alpha\dot{x})(-\dot{x}_p + \alpha(x_{ref} - x_p)). \tag{6.23}$$

Applying (5.18e) in (6.23), the criterion on stable piston motion can be defined by (6.16) as

$$K_1\dot{x}_p + K_2\dot{x}_p^2 + K_3\alpha(x_{ref} - x_p) + K_4\alpha(x_{ref} - x_p)\dot{x}_p < 0, \tag{6.24}$$

where the coefficients are

$$K_1 = \frac{P_1A_1}{m} - \frac{P_2A_2}{m}, \tag{6.25}$$

$$K_2 = -\frac{1}{m} + \alpha \tag{6.26}$$

$$K_3 = -\frac{P_1A_1}{m} + \frac{P_2A_1}{m}, \tag{6.27}$$

$$K_4 = \frac{1}{m} - \alpha. \tag{6.28}$$

The reference signal is tracked if (6.24) is ensured in the control algorithm. This means that the pressures P_1 and P_2 must be selected according to (6.24).

As a result, the following control algorithm is formulated.

1. Since P_1 and P_2 are interconnected by the value of u_{hyd}, the requirement on system stability is evaluated in each position of the valve $[A(x_{v,min}); 0; A(x_{v,max})]$, i.e., the calculation of (6.24) is requested.
2. Next, it is required to choose the value $A(x_v)$, where (6.24) achieves its minimum.
3. Moreover, the valve can be closed in some scenarios to provide energy-efficient control. If the condition on (6.24) and the inequality $\frac{x_{ref}-x_p}{x_{ref}} \cdot 100 < 1\%$ can be ensured at closed position of the valve, its opening is unnecessary.

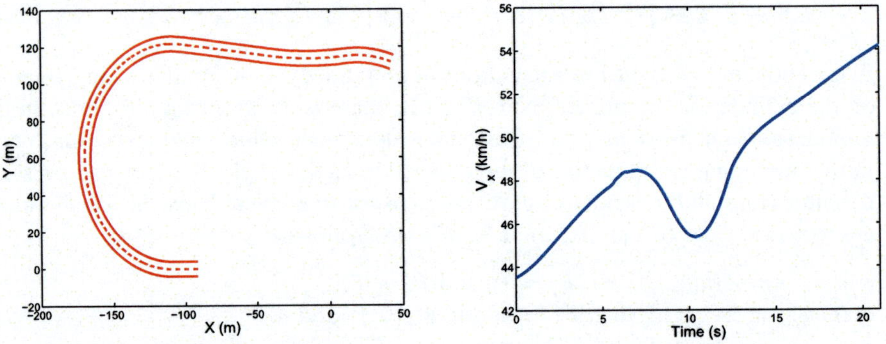

(a) Path for the vehicle (b) Velocity profile

Fig. 6.8 Simulation scenario for the illustration (©2017 IEEE. Reprinted, with permission, from Németh et al. 2017)

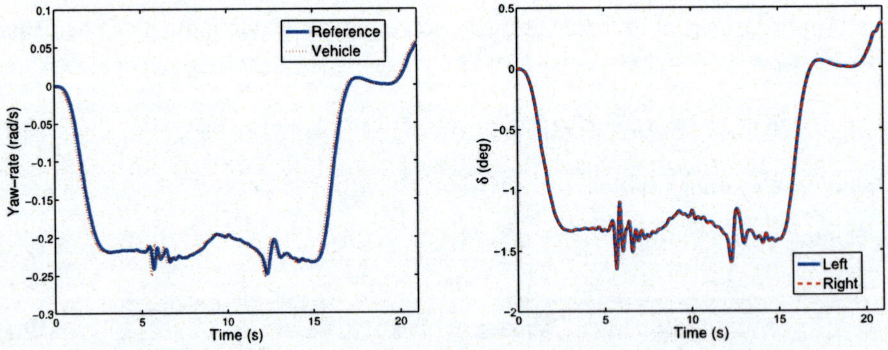

(a) Yaw-rate $\dot{\psi}$ (b) Steering angle δ

Fig. 6.9 Results of the simulation on vehicle performances (©2017 IEEE. Reprinted, with permission, from Németh et al. 2017)

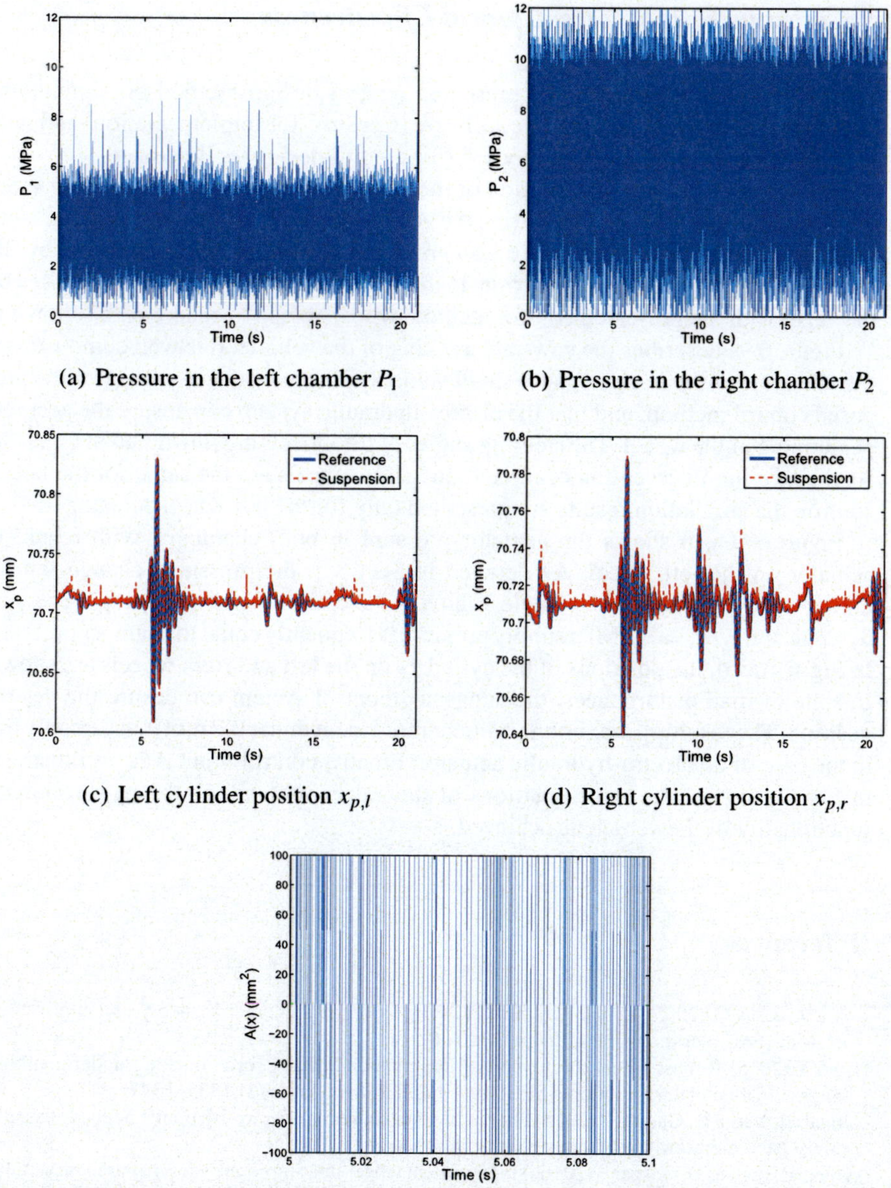

(a) Pressure in the left chamber P_1

(b) Pressure in the right chamber P_2

(c) Left cylinder position $x_{p,l}$

(d) Right cylinder position $x_{p,r}$

(e) Opening surface of the valve $A(x_v)$

Fig. 6.10 Results of the simulation on actuator performances (©2017 IEEE. Reprinted, with permission, from Németh et al. 2017)

6.4.2 *Illustration of the Control Effectiveness*

Finally, the effectiveness of the presented control design method on simulations with high-fidelity vehicle dynamic software (CarSim® for implementation) and with MATLAB® and Simscape Multibody™ (for control design) is illustrated.

The scenario provides the motion of the vehicle on a curvy race track (Fig. 6.8a) with varying speed (Fig. 6.8b). During the vehicle demonstration, a cornering maneuver has to be executed, in which the steering of the front axle is carried out by the variable-geometry vehicle suspension. In this simulation, the actuator is controlled by the valve with the presented control method, whose role is to ensure path following.

Figure 6.9a describes the yaw-rate tracking of the vehicle. It is well demonstrated that the error of the tracking has a small value. It illustrates the efficiency of the proposed control method, and that the electro-hydraulic system can ensure the accurate positioning of the wheel. The steering angles of the wheels are illustrated in Fig. 6.9b. As the steering intervention of the left and right wheels are the same for the lateral control, the simulation results are presented only for the left wheel actuator.

Figure 6.10a, b shows the actuator pressure in both chambers, with regard to actuator on the left wheel. As detailed in Sect. 8.2, the pressures change rapidly during the operation of the vehicle, which is the consequence of the rapid pressure dynamics, e.g., $P_s = 12\,\text{MPa}$ supply pressure is frequently equal to chamber pressure. In Fig. 6.10c, d, the positions of the cylinders on the left and right wheels are shown. In spite of road disturbances, the suggested control system can ensure the desired position. The resulting maximum cylinder movement under 0.2 mm is acceptably low in the case of an electro-hydraulic actuator. Finally, control input $A(x)$ is illustrated in Fig. 6.10e, i.e., the three positions of the valve, with which the path following functionality of the vehicle is achieved.

References

Bokor J, Balas G (2005) Linear parameter varying systems: a geometric theory and applications. In: 16th IFAC world congress, Prague, pp 1–6

Hsiao T (2015) Robust wheel torque control for traction/braking force tracking under combined longitudinal and lateral motion. IEEE Trans Intell Trans Syst 16(3):1335–1347

Németh B, Varga B, Gáspár P (2015) Hierarchical design of an electro-hydraulic actuator based on robust LPV methods. Int J Control 88(8):1429–1440

Németh B, Fényes D, Gáspár P (2016a) Independent wheel steering control design based on variable-geometry suspension. In: 8th IFAC international symposium on advances in automotive control, Norrköping, Sweden, pp 435–440

Németh B, Fényes D, Gáspár P, Bokor J (2016b) Trajectory tracking based on independently controlled variable-geometry suspension for in-wheel electric vehicles. In: 55th IEEE conference on decision and control, Las Vegas

Németh B, Fényes D, Gáspár P, Bokor J (2017) Control design of an electro-hydraulic actuator for variable-geometry suspension systems. In: 2017 25th Mediterranean conference on control and automation (MED), pp 180–185

Németh B, Fényes D, Gáspár P, Bokor J (2019) Coordination of independent steering and torque vectoring in a variable-geometry suspension system. IEEE Trans Control Syst Technol 27(5):2209–2220

Packard A, Balas G (1997) Theory and application of linear parameter varying control techniques. In: American control conference, workshop I, Albuquerque, New Mexico

Sename O, Gáspár P, Bokor J (2013) Robust control and linear parameter varying approaches. Springer, Berlin

Sontag E (1989) A "universal" construction of Artstein's theorem on nonlinear stabilization. Syst & Control Lett 13:117–123

Wu F, Yang XH, Packard A, Becker G (1996) Induced l^2-norm control for LPV systems with bounded parameter variation rates. Int J Nonlinear Robust Control 6:983–998

Wu FK, Yeh T, Huang CF (2013) Motor control and torque coordination of an electric vehicle actuated by two in-wheel motors. Mechatronics 23:46–60

Chapter 7
Coordinated Control Strategy for Variable-Geometry Suspension

In the former chapters, various controller designs have been introduced in a hierarchical structure. While the introduced control systems ensure performance and stability, effective coordination of torque vectoring and independent steering can be accomplished through the distribution of steering/force and reconfiguration of the actuators. This chapter suggests methods, which are formed through a polynomial analysis on variable-geometry vehicle suspension.

7.1 Motivations

Reconfiguration is motivated by achieving high efficiency through suspension control. Nevertheless, under some vehicle maneuvering situations, not all of the actuators are efficient. Longitudinal tracking/braking force on the wheel greatly affects steering intervention efficiency. For small $|F_{l,i}|$ force values, a small effect on the steering intervention through wheel tilting is yielded. Hence, in this case, vehicle motion is recommended to influence by torque vectoring. Avoiding ineffective actuation is ensured through the reconfiguration and the allocation of steering/force.

The problem behind the variation of suspension intervention effectiveness is as follows. During variable-geometry suspension control, wheel tilting γ_i can be calculated as

$$\gamma_i = \frac{u_i}{F_{l,i}}, \tag{7.1}$$

where u_i is the controller output, $F_{l,i}$ is the longitudinal control force of the wheel and i is the index associated with the right/left wheel. In case $F_{l,i}$ has a small value (e.g., near zero), the camber angle increases inadequately, such as

B. Németh and P. Gáspár, *Control of Variable-Geometry Vehicle Suspensions*, Advances in Industrial Control, https://doi.org/10.1007/978-3-031-30537-5_7

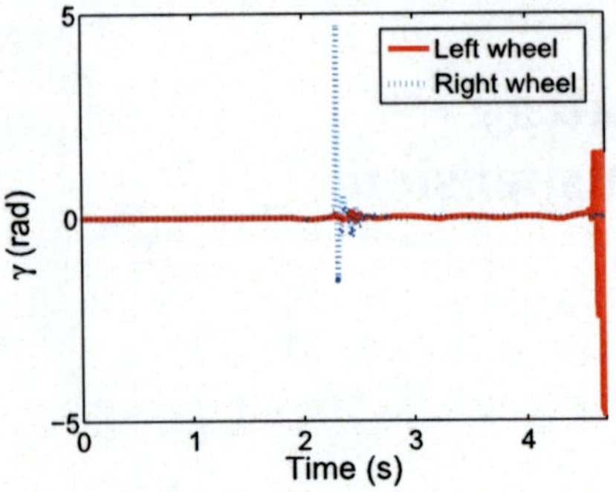

(a) Wheel cambers

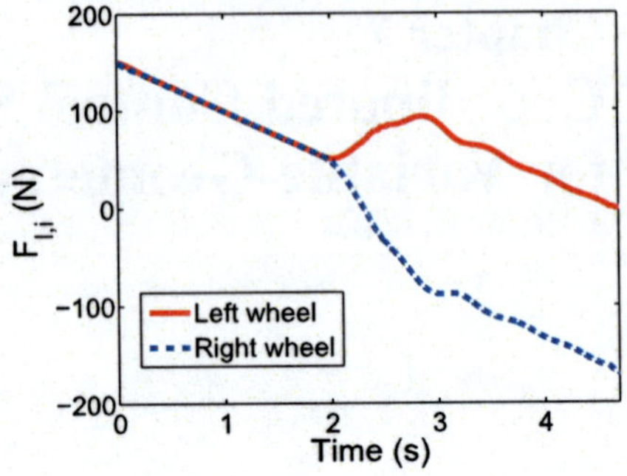

(b) Longitudinal forces

Fig. 7.1 Illustration for problem formulation (©2017 IEEE. Reprinted, with permission, from Fényes et al. 2016)

$$\lim_{F_{l,i} \to 0} \frac{F_{l,i}\gamma_i}{F_{l,i}} \Rightarrow \infty, \quad \gamma_i \Rightarrow \infty, \tag{7.2}$$

which leads to a zero-crossing problem, as illustrated in Fig. 7.1. Increasing γ_i can be disadvantageous, because the requested wheel tilting cannot be ensured due to limitations on the intervention capability, or sudden growth in γ_i leads to the significant reduction of the performance level on lateral vehicle dynamics.

A solution to the problem can be found through the reconfiguration of suspension intervention, i.e., at zero-crossing situations torque vectoring is prioritized. Providing priorities for the interventions requires carrying out examinations of their effectiveness, which is also introduced in this chapter.

7.2 Distribution Method of Steering and Forces on the Wheels

The scheme of the coordinated control, incorporating all of the control elements, can be seen in Fig. 7.2. The goal of the coordination is to select the corresponding signals $F_{l,i}$ and δ_l, δ_r, with which zero-crossing can be avoided. By choosing longitudinal forces, the ineffectiveness of the variable-geometry suspension should be avoided due to the low value of $|F_{l,i}|$. In addition, δ_l, δ_r have to be adjusted to the current longitudinal force ($F_{l,i}$). The coordination has been yielded by two steps, i.e., using M_d control torque the force $F_{l,i}$ is allocated and, then, δ_i is selected for each wheel.

Many studies propose strategies for optimal selection of $F_{l,i}$ values of each wheel; see, e.g., Ono et al. (2006), Sawase and Ushiroda (2008) and Klomp (2011). Since $F_{l,i}$ has an impact on the longitudinal dynamics as well, the sum of $F_{l,i}$ has to be equal to the requested cumulative longitudinal traction force F_0, which is provided

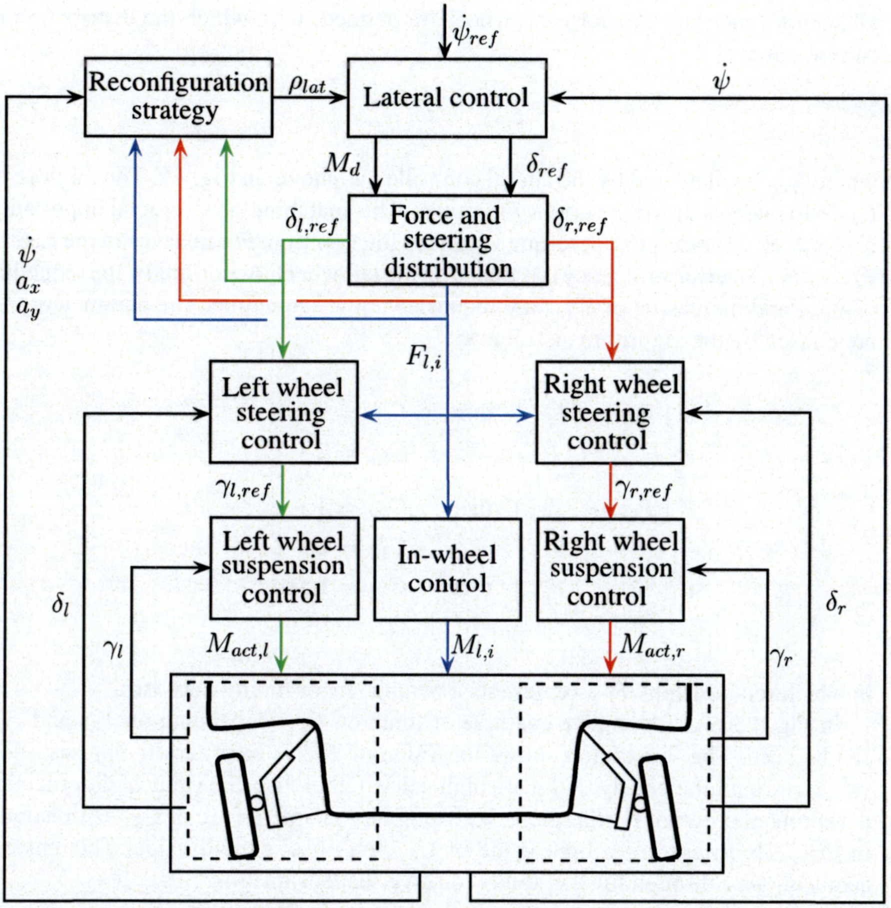

Fig. 7.2 Scheme of the coordinated control system architecture (©2018 IEEE. Reprinted, with permission, from Németh et al. 2019)

by the cruise control system for achieving speed tracking functionality. Here, F_0 is assumed to be known, and a constraint on the selection of wheel forces is not to modify F_0 due to the distribution. Accordingly, the traction/braking forces on both sides are computed as

$$F_{l,l} = F_0 + \frac{M_d}{b_f}, \quad F_{l,r} = F_0 - \frac{M_d}{b_f}, \tag{7.3}$$

in which expressions b_f denotes the length of the front vehicle track.

The selection of traction/braking forces requires also the selection of reference steering angles $\delta_{i,ref}$ for each wheel. In the lateral control design process, the average of $\delta_{i,ref}$ is considered, i.e., δ_{ref} is produced by the lateral control. For steering angle allocation functions $C_{\delta,i}$ for each wheel are defined, with which the distribution is carried out as

$$\delta_{i,ref} = C_{\delta,i}\delta_{ref}, \tag{7.4}$$

where δ_{ref} is calculated by the lateral controller, as shown in Fig. 7.2. The purpose of $C_{\delta,i}$ is to match δ_{ref} to the current $F_{l,i}$ values. This matching plays crucial importance if $F_{l,l} \neq F_{l,r}$, for example at torque vectoring intervention. For instance in the case of $F_{l,r} > F_{l,l}$, steering efficiency is reduced on the left wheel. Accordingly, the reduction of $\delta_{l,ref}$ and increasing of $\delta_{r,ref}$ are requested at the same time. The parameters $C_{\delta,i}$ are chosen by the algorithm as follows:

$$\begin{aligned}
&\text{if } \mid F_{l,l} \mid < \mid F_{l,r} \mid, \text{ then} && C_{\delta,l} = 1 - e^{-\frac{F_{lim}}{|F_{l,r}|-|F_{l,l}|}} \\
& && C_{\delta,r} = 2 - C_{\delta,l} \\
&\text{if } \mid F_{l,r} \mid < \mid F_{l,l} \mid, \text{ then} && C_{\delta,r} = 1 - e^{-\frac{F_{lim}}{|F_{l,l}|-|F_{l,r}|}} \\
& && C_{\delta,l} = 2 - C_{\delta,r} \\
&\text{if } \mid F_{l,l} \mid = \mid F_{l,r} \mid, \text{ then} && C_{\delta,l} = 1 \text{ and } C_{\delta,r} = 1,
\end{aligned} \tag{7.5}$$

in which relationships F_{lim} represents a parameter for tuning purposes.

In Fig. 7.3a, an illustrative example of function $C_{\delta,l}$, depending on $F_{l,r}$ and $F_{l,l}$, can be seen. The illustration shows the value of $C_{\delta,l}$ is near zero if $F_{l,l}$ has a low value, and simultaneously, $F_{l,r}$ has a high value. Otherwise, the value of $C_{\delta,l}$ function is near its maximum. Furthermore, an exemplary characteristic of $C_{\delta,r}$ is illustrated in Fig. 7.3b. In case of a high value of $C_{\delta,r}$, $C_{\delta,l}$ has a small value. This ensures achieving δ_{ref} through the variable-geometry suspension.

7.3 Reconfiguration Strategy

The reconfiguration strategy has been designed to provide a balance between torque vectoring and suspension-based steering functionalities. In the proposed method, the effectiveness of the interventions is highlighted, which is based on a polynomial analysis of the lateral vehicle dynamics, especially on a reachability set analysis. These results are incorporated into the reconfiguration strategy, also considering the effect of longitudinal forces on steering ability. This creates a strategy for actuation selection between torque vectoring and steering, in which the reconfiguration is represented by scheduling variable ρ_{lat}.

Fig. 7.3 Example of the selection of functions for steering angle distribution (©2018 IEEE. Reprinted, with permission, from Németh et al. 2019)

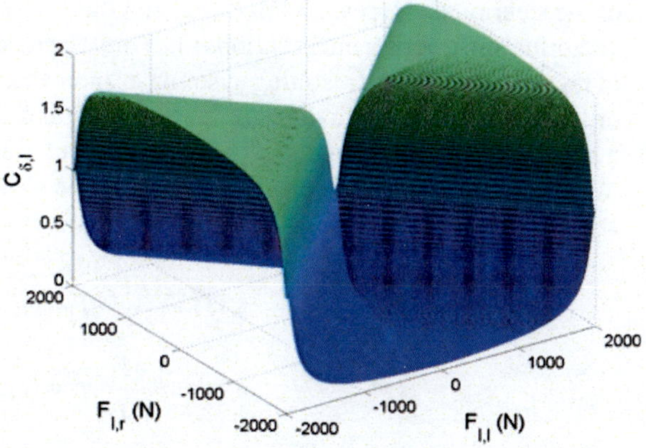

(a) Function of $C_{\delta,l}$

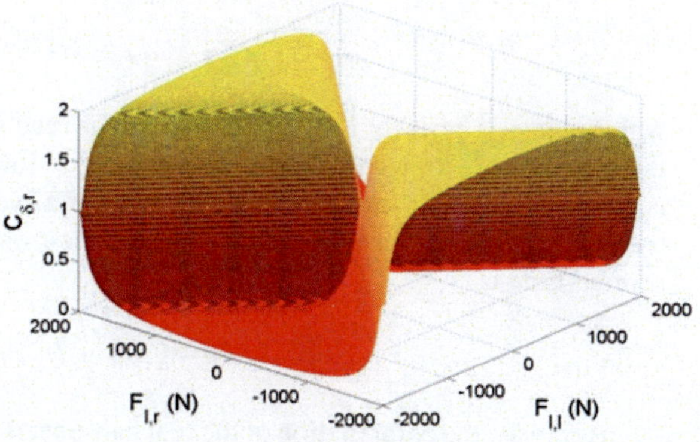

(b) Function of $C_{\delta,r}$

The reachable set of the states can be defined as follows; see Boyd et al. (1997). It is given system $\dot{x} = f(x(t)) + gu(t)$ in continuous time, which has $x(0) = 0$ initial condition. The reachable set under the condition of peak-bounded inputs is

$$\mathscr{R} \triangleq \left\{ x(T) \left| \begin{array}{l} (x(t), M_d(t), \Delta(t)) \,| \\ \dot{x}(t) = f(x(t)) + gM_d(t) + h\Delta(t), x(0) = 0, \\ \quad \Delta_{min} \le \Delta(t) \le \Delta_{max}, \\ \quad M_{d,min} \le M_d(t) \le M_{d,max}, T \ge 0 \end{array} \right. \right\}. \tag{7.6}$$

The defined $\mathscr{R}$ reachable set involves states of the system, which are reachable by the bounded control inputs, such as through torque vectoring and/or steering.

In the analysis of variable-geometry vehicle suspension intervention, the dynamics of the steering angle realization (5.14) and the lateral vehicle dynamics under torque vectoring and steering interventions (4.7) must be incorporated. Due to the connection of these dynamics, the resulting system has increased state number. Increased system order may result in difficulties in the analysis, and, therefore, it is recommended to avoid it through a separate reachable set computation.

Dynamics of wheel steering is formulated in (5.14), and, thus, solutions of δ_i, $\dot{\delta}_i$ signals are as follows:

$$\dot{\delta}_i(t) = \frac{\varepsilon F_{l,i}}{J_{\delta,i}} t \gamma_i(t), \tag{7.7a}$$

$$\delta_i(t) = \frac{\varepsilon F_{l,i}}{2J_{\delta,i}} t^2 \gamma_i(t), \tag{7.7b}$$

where $F_{l,i}$ is considered to have a fix value. Furthermore, during the analysis time range is considered to be limited to $t = T$, i.e., reachable set on $\dot{\delta}_i(T), \delta_i(T)$ is computed. Accordingly, (4.7) can be reordered as

$$\dot{x} = f(x) + gM_d + H(F_{l,i}, T)\gamma, \tag{7.8}$$

which creates a basis for the computation of the reachable set. It is carried out through the procedure of trajectory reversing; see, e.g., Horiuchi (2015). In this process, it is considered that the null-controllability set of a system in forward-time is equal to the reachable set of the same system in reverse-time Snow (1967). System (7.8) in reverse-time is

$$\dot{x} = -f(x) - gM_d - H(F_{l,i}, T)\gamma. \tag{7.9}$$

In this way, the computation of the reachable set is converted to a controllability set computation problem. Paper Németh et al. (2016) suggests a Sum-of-Squares (SOS) programming-based method to approximate the controllability set of polynomial systems, such as the system of (7.9). The calculation method of the set demands the existence of a positive-definite, proper and smooth $V : \mathbb{R}^n \rightarrow \mathbb{R}$ Control Lyapunov Function, which requires that $\inf_{u \in \mathbb{R}} \left\{ \frac{\partial V}{\partial x}(-f(x)) + \frac{\partial V}{\partial x}(-M) \cdot u \right\} < 0$ must be guaranteed for all $x \neq 0$, where $M = \begin{bmatrix} g & H \end{bmatrix}$ and $u = \begin{bmatrix} M_d & \gamma \end{bmatrix}^T$.

Accordingly, the optimization task for the approximation of the maximum Controlled Invariant Set is formed as

$$\max \beta \tag{7.10}$$

over polynomials $V, p_1, p_2 \in \mathscr{R}_n$, $V(0) = 0$ and SOS polynomials $s_1, s_2, s_3, s_4, s_5 \in \Sigma_n$, subject to the constraints

$$-\left(-\frac{\partial V}{\partial x}f(x)+\frac{\partial V}{\partial x}g\,u_{max}\right)-s_1\left(-\frac{\partial V}{\partial x}g-\varepsilon\right)-s_2\,(1-V)-p_1L_1\in\Sigma_n \tag{7.11a}$$

$$-\left(-\frac{\partial V}{\partial x}f(x)-\frac{\partial V}{\partial x}g\,u_{max}\right)-s_3\left(\frac{\partial V}{\partial x}g-\varepsilon\right)-s_4\,(1-V)-p_2L_2\in\Sigma_n \tag{7.11b}$$

$$-\,(s_5(\beta-p)+(V-1))\in\Sigma_n. \tag{7.11c}$$

In the optimization problem, β denotes the level set of $P_\beta := \{x \in \mathbb{R}^n | \; p(x) \le \beta\}$, $p \in \Sigma_n$ is a fixed and positive-definite function, $L_{1,2}(x)$ is a positive-definite polynomial and $\varepsilon \in \mathbb{R}^+$ is a selected constant with small value. The optimization must be performed for both interventions, such as torque vectoring and wheel tilting.

Reachable sets with steering, i.e., tilting intervention with its bound $\gamma = \pm 1.5°$ at various F_l can be seen in Fig. 7.4. The characteristics of the computed sets show that the reachable states significantly depend on F_l. For example, at $F_l = 0$ reachable states are not found and thus, wheel tilting intervention is inefficient. The analysis confirms that it can be critical to reconfigure the actuation of the camber angle at varying F_l.

Illustration of the reachable sets with torque vectoring, depending on the selection of different maximum values on M_d is found in Fig. 7.5. The shape of these reachable sets differs from the shape of the reachable set in (7.4), i.e., wheel tilting intervention is more efficient for α_1 variation, while torque vectoring has higher efficiency for α_2

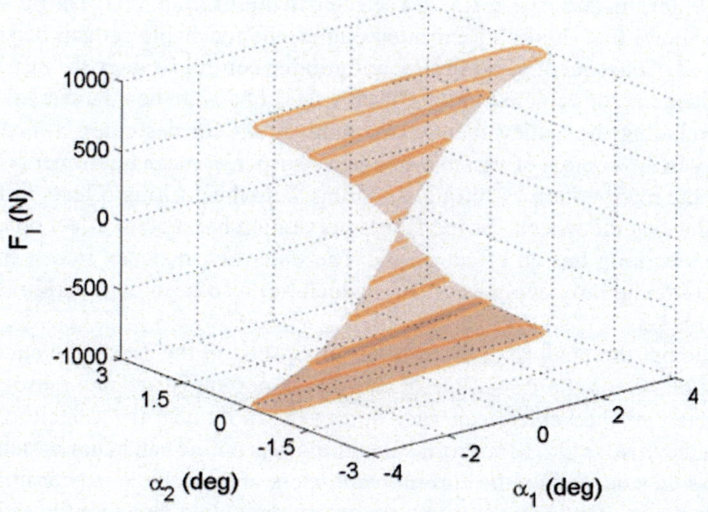

Fig. 7.4 Illustration of reachable sets with steering intervention (©2018 IEEE. Reprinted, with permission, from Németh et al. 2019)

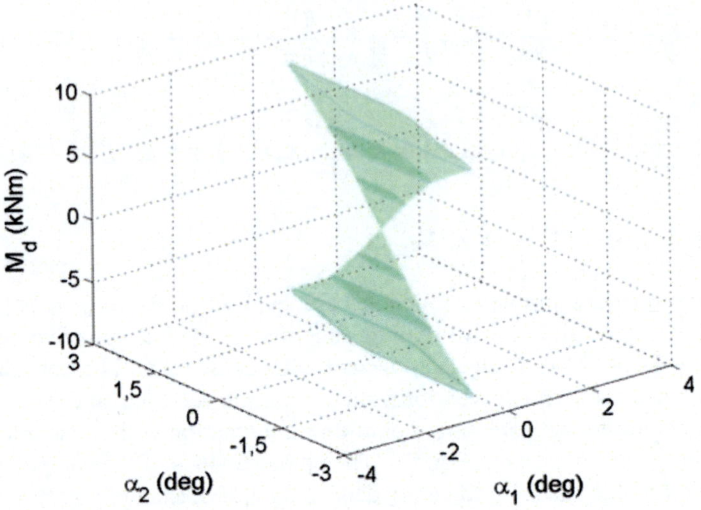

Fig. 7.5 Illustration of reachable sets with torque vectoring intervention (©2018 IEEE. Reprinted, with permission, from Németh et al. 2019)

variation. Differences in the characteristics highlight the benefit of reconfiguration, such as the possibility of reaching different states.

Moreover, reachability analysis is considered for the joint intervention of torque vectoring and wheel tilting, as Fig. 7.6 shows. Maximum values on the interventions $M_d = 9000$ Nm and $\gamma = \pm 1.5°$ are selected in this examination. The presented illustration shows that through joint intervention, the reachable set can be significantly enhanced. Consequently, using a reconfigurable control strategy the effectiveness of the influencing of vehicle motion for all values of F_l can be guaranteed.

Concluding the achievements, two motivations for designing a reconfiguration strategy in the context of variable-geometry suspension can be formed. One of them is that the intervention of torque vectoring and wheel tilting affects different side-slip values on the wheels. Suspension intervention has a great effect on α_1, whereas torque vectoring has an effect on α_2. The other consequence is that the effect of wheel tilting greatly depends on $F_{l,i}$, which has also been explored in the shape of reachable sets.

Although the analysis has led to accurate sets on the reachable states, in practical applications the direct use of the sets is not suggested. Its reason is that the parameters of the vehicle can vary during the motion of the vehicle, for example the characteristics $\mathscr{F}_i$. Moreover, calculating sets online can be numerically complicated as they depend on the current parameters of the vehicle. Accordingly, the two previously developed implications are incorporated into the reconfiguration strategy design. This leads to the next reconfiguration rules (Németh et al. 2019).

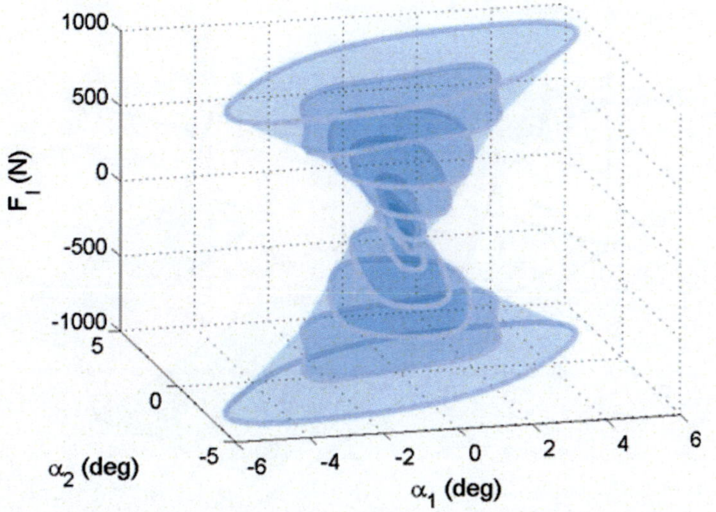

Fig. 7.6 Illustration of reachable sets with joint intervention (©2018 IEEE. Reprinted, with permission, from Németh et al. 2019)

- Since torque vectoring actuation can result in significant power loss (Gruber et al. 2016), steering intervention is preferable. Accordingly, the steering operates at $\rho_{lat,min}$ during normal cruising for economic reasons.
- In case the tire slip α_2 grows, pure steering intervention is not effective as it is shown in Fig. 7.4. Hence, as α_2 increases, the value of ρ_{lat} can also be increased. This results in increased system intervention through joint actuation as Fig. 7.6 shows.
- Since $F_{l,i}$ has a high impact on the reachability of the steering with variable-geometry suspension, the cutback of longitudinal forces on the wheel through the reconfiguration can be achieved. Accordingly, a reduction of $F_{l,i}$ can lead to growing ρ_{lat}.

The rules of the reconfiguration strategy can be formed in a function, which is defined as

$$\rho_{lat} = \mathfrak{f}(\alpha_2, F_{l,r}, F_{l,l}). \tag{7.12}$$

The shape of $\mathfrak{f}$ is based on the rules above; see an illustration in Fig. 7.7. Variation of ρ_{lat} between $\rho_{lat,min}$ and $\rho_{lat,max}$ can be performed. In the case of low ρ_{lat} steering is preferred, but at high ρ_{lat} torque vectoring has priority. If α_2 achieves a high value, the selection of $\rho_{lat,int}$ leads to integration, and the reduction of $F_{l,i}$ results in the increase of ρ_{lat}. Forming the shape of $\mathfrak{f}$ requires the selection of $p_{1,2,3,4}$ design parameters; see Fig. 7.7. The role of distinctions between p_1, p_2 and also p_3, p_4 is the avoidance of the chattering of control inputs. The parameter selection can be yielded by using Monte Carlo simulations, with which the effect of transi-

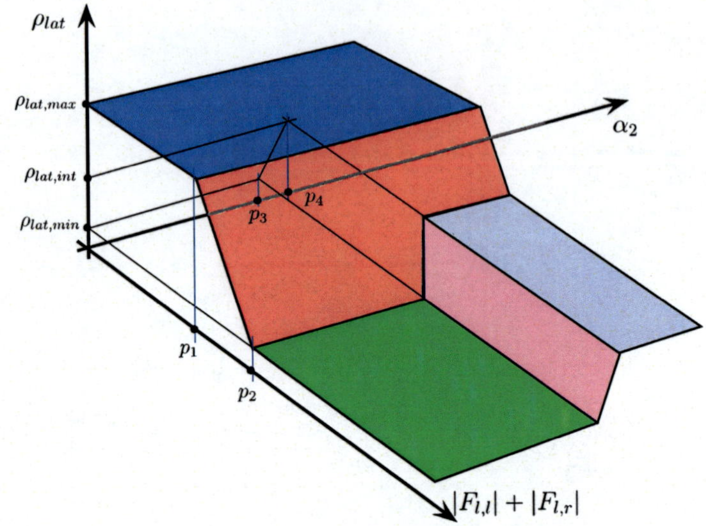

Fig. 7.7 Selection of ρ_{lat} value based on $\mathfrak{f}$ (©2018 IEEE. Reprinted, with permission, from Németh et al. 2019)

tions of the control inputs on z_{lat} must be minimized. Accordingly, many simulations have been evaluated in which F_0 and y_{ref} variations have been performed. First, the controllers $K_{lat}(\rho_{lat})$, $K_{st,i}(\rho_{st,i})$ are designed and for forming $\mathfrak{f}$, initial values of $p_{1,2,3,4}$ are selected. Second, a simulation with a variable value of F_0, y_{ref} is evaluated for achieving reconfiguration between the control inputs. Third, the parameters p_1, p_2, p_3, p_4 based on the evaluation of the simulations are modified for minimizing z_{lat} (Gill et al. 1981). The reconfiguration method demands the values $F_{l,i}$ supplied by the force distribution block, as shown in Fig. 7.2. Furthermore, an estimate of α_2 is needed by measuring lateral and longitudinal acceleration, steering wheel angle, wheel angular velocities and yaw rate; see Grip et al. (2008).

7.4 Illustration of the Reconfiguration Strategy

Finally, in this chapter a simulation in CarSim® on the reconfigurable control system for illustrating its efficiency is presented. A D-class passenger car with 1690kg in the simulations is used, which has been driven along on Melbourne Grand Prix course, as is depicted in Fig. 7.8. The length of the course is around 5000 m and includes many curves. Parameters for forming $\mathfrak{f}(\alpha_2, F_{l,r}, F_{l,l})$ are selected as $p_1 = 500N$, $p_2 = 200N$, $p_3 = 0.3°$ and $p_4 = 0.5°$.

The longitudinal velocity of the vehicle can be seen in Fig. 7.9a. The resulting longitudinal forces of the wheels, due to accelerations and decelerations, are shown in Fig. 7.9b. The trajectory tracking is adequately precise, with the maximum lateral

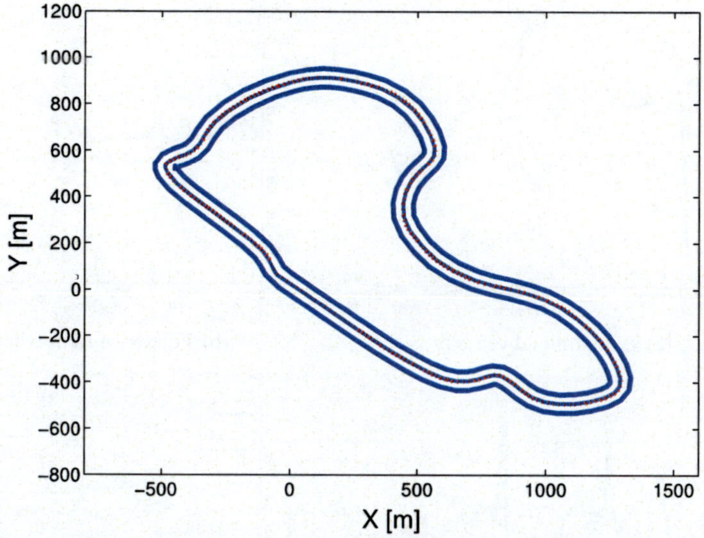

Fig. 7.8 Track of the vehicle during the simulation (©2018 IEEE. Reprinted, with permission, from Németh et al. 2019)

error being approximately 0.18 m, as is shown in Fig. 7.9c. Figure 7.9d,e illustrates steering intervention and M_d for torque vectoring. The trajectories of the vehicle lateral states together with the computed reachable sets, in the function of α_1, α_2, F_l, can be seen in Fig. 7.9f. During the entire simulation scenario, the state trajectory is inside of the reachable sets, which confirms the successful approximation of the sets.

Illustrations of details of the intervention of variable-geometry vehicle suspension-based steering in Fig. 7.10 can be found. The value of the left wheel tilting, which results in the steering angle on the left wheel, can be seen in Fig. 7.10a. The detail on the low-level suspension intervention, such as actuator torque, is shown in Fig. 7.9b, which between −300 Nm ... 300 Nm physical limits varies. During the simulation, simplified actuator model with second-order dynamics has been applied; see Németh et al. (2017). Finally, Fig. 7.10c illustrates ρ_{lat} and rear wheel slip in Fig. 7.10d. The results show that longitudinal wheel forces at road section 2900 m ... 3000 m have high values, which induces high γ_i intervention, although α_2 is also large, indicating that through pure steering low lateral error cannot be guaranteed. Thus, the steering has been reconfigured to a joint intervention through the selection of ρ_{lat}, around 0.5. Consequently, through the reconfiguration-based coordination of torque vectoring and variable-geometry vehicle suspension intervention, the safe and efficient motion of the vehicle can be achieved.

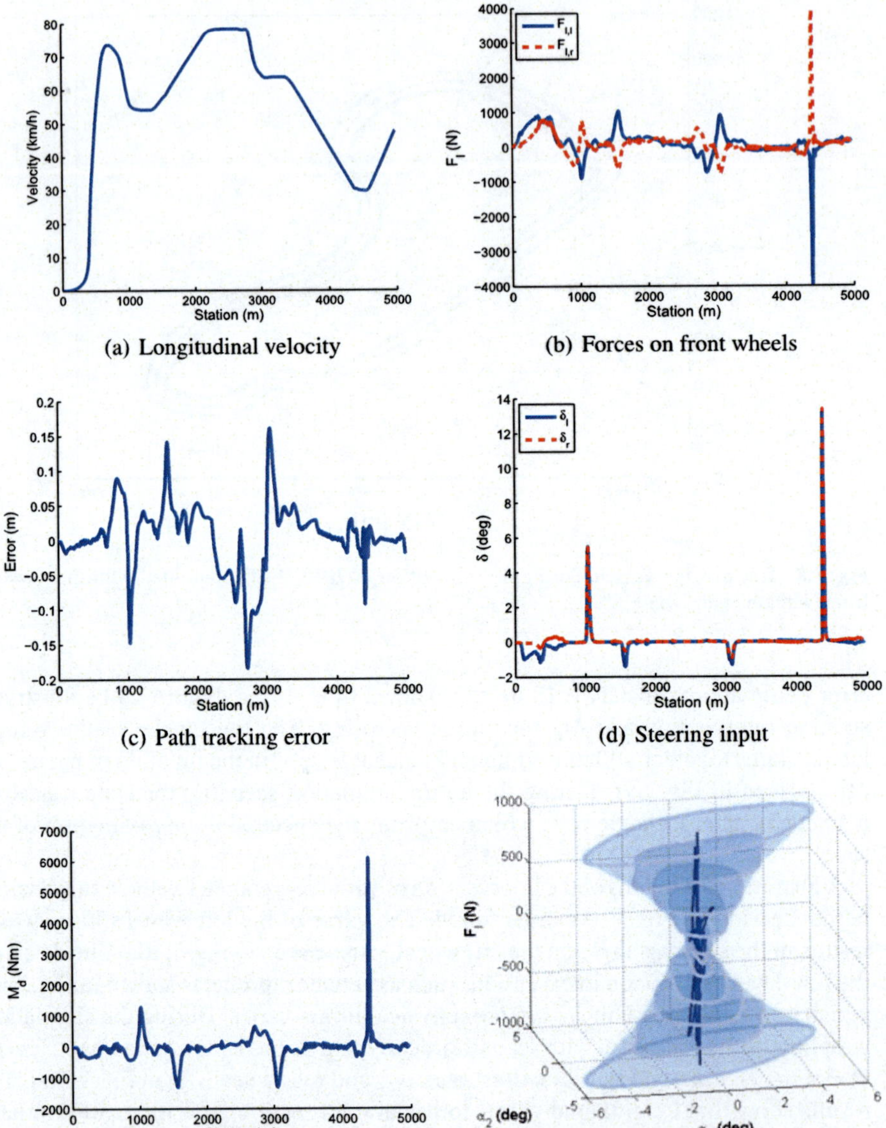

(a) Longitudinal velocity
(b) Forces on front wheels
(c) Path tracking error
(d) Steering input
(e) Torque vectoring input
(f) Reachable sets and state trajectory

Fig. 7.9 Simulation results on path following (©2018 IEEE. Reprinted, with permission, from Németh et al. 2019)

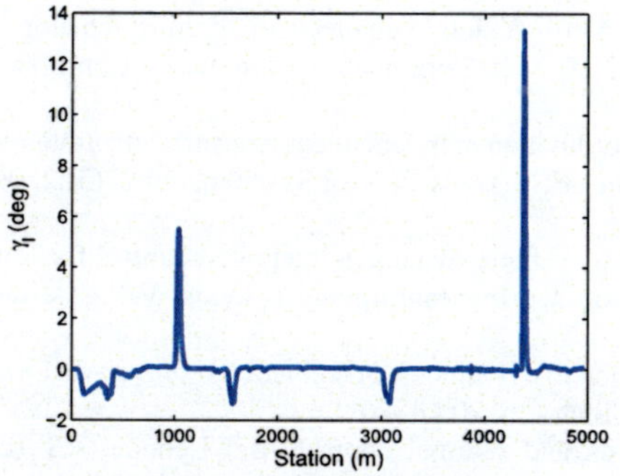

(a) Left wheel tilting

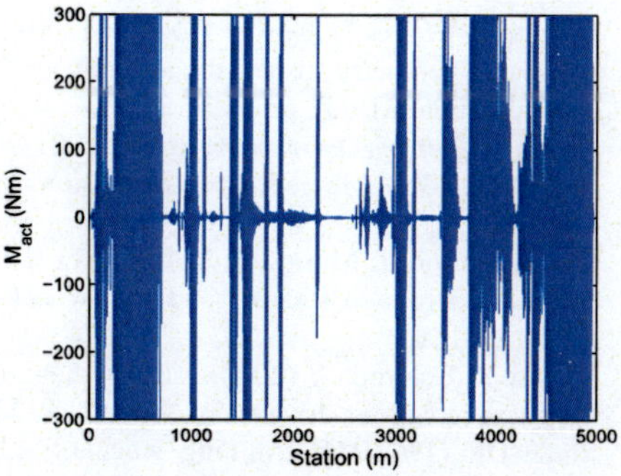

(b) Torque actuation on the left wheel

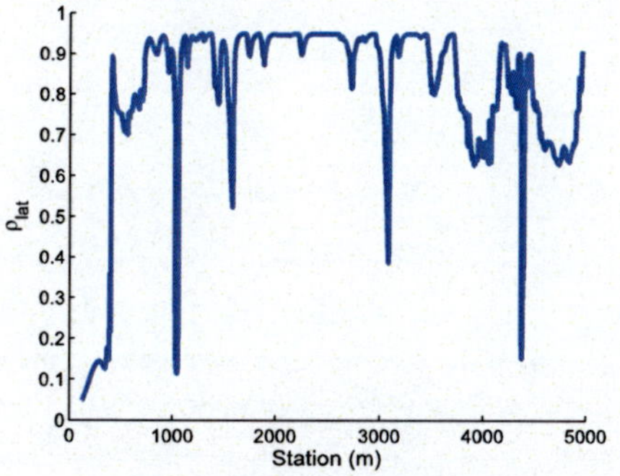

(c) Scheduling variable

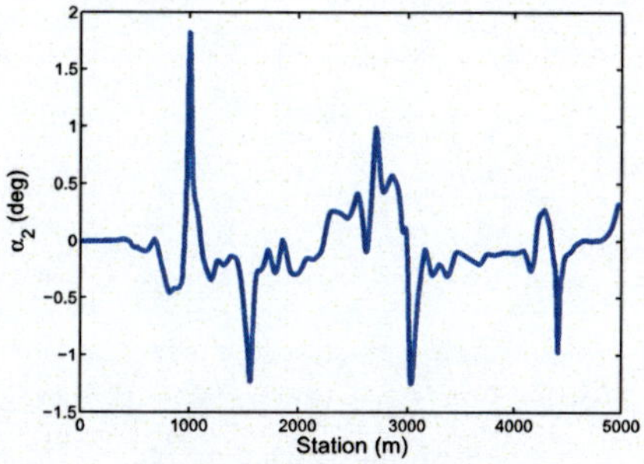

(d) Rear wheel side slip

Fig. 7.10 Simulation results on low-level operation (©2018 IEEE. Reprinted, with permission, from Németh et al. 2019)

References

Boyd S, Ghaoui LE, Feron E, Balakrishnan V (1997) Linear matrix inequalities in system and control theory. Society for Industrial and Applied Mathematics, Philadelphia

Fényes D, Németh B, Gáspár P (2016) Handling of zero-crossing problems in the design of variable-geometry suspension control. In: 2016 IEEE 17th International Symposium on Computational Intelligence and Informatics (CINTI). pp 237–242

Gill PE, Murray W, Wright M (1981) Practical optimization. Academic Press, London UK

Grip H, Imsland L, Johansen T, Fossen T, Kalkkuhl J, Suissa A (2008) Nonlinear vehicle side-slip estimation with friction adaptation. Autom 44(11):611–622

Gruber P, Sorniotti A, Lenzo B, Filippis GD, Fallah S (2016) Energy efficient torque vectoring control. In: Edelmann J, Plöchl M, Pfeiffer PE (eds) Advanced vehicle control AVEC'16. CRC Press/Balkema, pp 17–22

Horiuchi S (2015) Evaluation of chassis control algorithms using controllability region analysis. In: International Symposium on Dynamics of Road Vehicles on Roads and Tracks. Graz, Austria

Klomp M (2011) Longitudinal force distribution using quadratically constrained linear programming. Veh Syst Dyn 49(12):1823–1836

Németh B, Gáspár P, Péni T (2016) Nonlinear analysis of vehicle control actuations based on controlled invariant sets. Int J Appl Math Comput Sci 26(1)

Németh B, Fényes D, Gáspár P, Bokor J (2017) Control design of an electro-hydraulic actuator for variable-geometry suspension systems. In: 2017 25th Mediterranean Conference on Control and Automation (MED). pp 180–185

Németh B, Fényes D, Gáspár P, Bokor J (2019) Coordination of independent steering and torque vectoring in a variable-geometry suspension system. IEEE Trans Control Syst Technol 27(5):2209–2220

Ono E, Hattori Y, Muragishi Y, Koibuchi K (2006) Vehicle dynamics integrated control for four-wheel-distributed steering and four-wheel-distributed traction/braking systems. Veh Syst Dyn 44:139–151

Sawase K, Ushiroda Y (2008) Improvement of vehicle dynamics by right and left torque vectoring system in various drive trains. Mitsubishi Mot Tech Rev 20:14–20

Snow DR (1967) Determining reachable regions and optimal controls. In: Leondes CT (ed) Advances in control systems, vol 5. Academic Press

Chapter 8
Control Implementation on Suspension Test Bed

This chapter introduces the implementation of steering control on variable-geometry vehicle suspension, using a one-wheel test bed, which is integrated with a Hardware-in-the-Loop (HiL) framework. A hierarchical control algorithm has been created as a contribution of this chapter, the efficiency of which has been shown during implementation. Control systems on low and high levels have been involved in the hierarchical control structure. A high-level controller has a role in trajectory following, whose functionality has been evaluated in a CarMaker high-fidelity simulator. A low-level control operation, i.e., generating steering angle, using a linear actuator in the test bed has been performed. The present chapter shows that the introduced approach of variable-geometry suspension control is capable of realizing steering functions.

8.1 Introduction to Test Bed for Variable-Geometry Vehicle Suspension

The variable-geometry vehicle suspension test bed used in this section is introduced, i.e., construction units and control architecture within the applied electronic architecture.

8.1.1 Test Bed Construction

The concept of the test bed in its physical realization can be seen in Fig. 8.1. The provided construction has only one main difference from the solution in Part II. Actuator motion in the vertical direction, connected to the test bed frame, can be achieved, while in the original concept the actuator is linked to the hub. The reason is that in a laboratory environment using a linear electric actuator is easier, instead

B. Németh and P. Gáspár, *Control of Variable-Geometry Vehicle Suspensions*, Advances in Industrial Control, https://doi.org/10.1007/978-3-031-30537-5_8

Fig. 8.1 View of test bed construction

of a hydraulic unit. Nevertheless, it leads to similar functionality, i.e., the operation of the suspension is the same (Fényes et al. 2022).

The test bed contains the next units:

I The DC motor has the role to provide driving torque on the wheel. It has an Actobotics 638280 type, which provides the proper wheel torque using a built-in planetary gearbox with a 121/3249 gear ratio. The unit is able to generate 3 Nm maximum torque, and its maximum rotational speed is 313 RPM. The requested voltage supply is between 6 ... 12 V. For measuring driving speed, the motor has an encoder on its shaft with 48 teeth on the gear.

II The linear actuator aims the achieving of wheel tilting. The type of the built-in actuator is LATT 4A 1/12 with 165 mm stroke and 12 V supply. Positioning of the actuator through a hall sensor-based encoder is realized, with which the maximum stroke can be set. An electronic board to the actuator has been connected, which provides current for actuator positioning. The actuator has been moved with constant speed, whose maximum is 37 mm/s, and it is available for push and pull motions.

III The lower suspension arm through a joint to the frame is connected, and, moreover, the other end connects to the hub. The arm has been linked to the frame with a one-dimensional joint and to the hub with a three-dimensional joint. There is another encoder on the lower arm to define the steering angle. In fact, it measures axis joint rotation. The voltage supply of the encoder is 5V, and its resolution is 3 ... 480 Lines Per Inch, i.e., 0.18°.

IV The upper suspension arm is a substantial unit in the McPherson suspension construction. The spring on the strut provides vertical load transmission, which is necessary to create traction force. The strut through a joint to the frame is connected, and, moreover, on its other end connected to the hub.

V The diameter size of the wheel is 18″, i.e., a wheel of a small-sized bicycle is applied. It is fixed to the hub shaft through a rigid connection, and, thus, its rotation speed is the same as the speed of the motor. It is able to be rotated in forward and backward directions.

VI The hub connects the DC motor with the wheel, i.e., the right side of the shaft is linked to the wheel, and the left side to the driving electric motor. Inside the hub, a special bearing construction together with a clutch can be found (for details, see Fényes et al. 2022). The role of the clutch is to compensate the mechanical eccentricity of the motor and hub shafts. Moreover, hub housing has a connection to both suspension arms and to the actuator.

VII A rotating disk is located under the wheel. It is able to rotate around its vertical axis, with which a surface for displacement and rotation of the wheel is provided.

The presented test bed has some limitations, which result from its control-oriented purpose, i.e., its purpose is to demonstrate the effectiveness of the control operation.

- The range of wheel steering angle has limits, which is caused by limits on achievable tilting between $-20^\circ \ldots 20^\circ$ and on wheel speed.
- The limitation on wheel speed results from the construction of the in-door test bed. The maximum available speed is 27 km/h. Its reason is the limited rotational speed of the driving motor.
- The linear actuator has an operating speed of 37 mm/s, which limits the vehicle maneuvers that can be performed. Furthermore, the set of executable maneuvers are limited due to time delay in the operation of the test bed.
- Although the wheel has stiffness, it can be considered rigid due to low vertical load. Thus, the value of the pneumatic trail and its variation can be neglected.
- Finally, due to the flat rotating plate, the roughness of the road during the tests cannot be considered.

8.1.2 Control Architecture in Human-in-the-Loop Simulations

A Hardware-In-the-Loop (HIL) setup has been created for the purpose of experimenting with the introduced test bed. The dynamics of the entire vehicle is modeled by the high-fidelity simulation software, CarMaker. The installation design includes additional elements described in Fig. 8.2. The main elements of the setup are given as follows. The PC has double purposes during the simulations. One of them is that it includes a simulation software system, such as MATLAB® and CarMaker. CarMaker has been used for modeling vehicle dynamics (longitudinal, vertical and lateral), together with the dynamics of steering. The measured signals on the test bed, which are inputs for CarMaker, are δ and v_x. The goal of CarMaker is to calculate

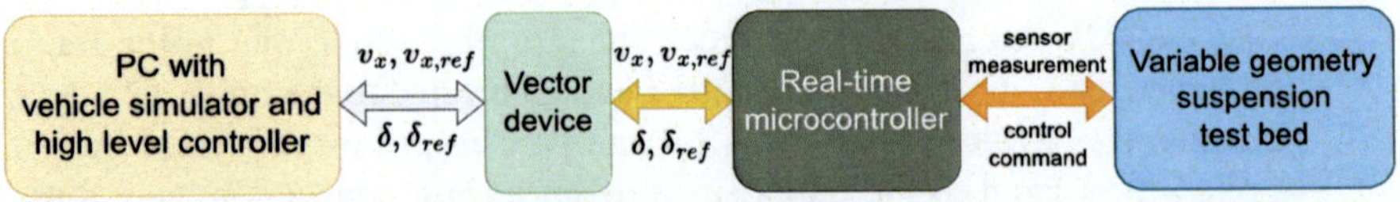

Fig. 8.2 Scheme of control system in the HiL simulation

lateral error y_e and heading error ψ_e of the vehicle; the sources of their computation are the actual vehicle states and the waypoints of the path. From CarMaker to MATLAB®, the signals y_e, ψ_e are transmitted. Another purpose of the software is to implement the high-level controller, especially under a MATLAB® environment. The outputs of the high-level control are two reference values for the low-level control system using CAN and serial communication protocols, involving in Vector device. The Vector device interface is used to transmit measured signals between sensors and MATLAB® and CarMaker on the PC. The calculated references to a real-time Microcontroller, i.e., dSPACE AutoBox II are transmitted, which device has various purposes in the control loop. AutoBox II device serves in the processing of various tasks. It processes raw measured signals of the sensors (v_x, δ), which are located on the test bed, such as wheel speed sensor, actuator positioning sensor and wheel tilting angle sensor. Moreover, AutoBox II manages data transfer with the PC and Vector device, using CAN and serial communications. Finally, one of the most important purposes of its application is to implement low-level control. It calculates the requested camber angle, linear actuator intervention and driving command of the DC motor from the signals v_x, $v_{x,ref}$, δ, δ_{ref}, i.e., PWM control signals to the test bed.

8.2 Implemented Control Algorithm on the Suspension Test Bed

A hierarchical solution has been used for the implementation of the control, which is illustrated in Fig. 8.2. The low-level control on AutoBox II has the responsibility for realizing $v_{x,ref}$, δ_{ref}, which are computed by the high-level control. High-level control calculates the references, using lateral and heading errors from CarMaker and from the test bed measurements.

The purpose of this chapter primarily is to demonstrate the efficiency and operation of the implemented variable-geometry vehicle suspension solution, and, thus, steering control functionality is evaluated. Consequently, $v_{x,ref}$ is constant during the test scenarios, whose tracking through a PDI controller on the low level is yielded; see, e.g., Sahputro et al. (2017) and Chamraz and Balogh (2018) methods on their design process.

8.2.1 Design on the High Level for Lateral Control Purposes

Lateral control is designed for ensuring path following. During the implementation process, a limited complexity model is used for control design, i.e., a one-wheel bicycle model, which includes equations on lateral and yaw motions:

$$mv_x(\dot{\psi} + \dot{\beta}) = C_1\alpha_1 + C_2\alpha_2, \tag{8.1a}$$

$$J\ddot{\psi} = C_1\alpha_1 l_1 - C_2\alpha_2 l_2, \tag{8.1b}$$

$$\dot{v}_y = v_x(\dot{\psi} + \dot{\beta}), \tag{8.1c}$$

in which representation J is inertia on yaw motion, m denotes vehicle mass, l_i values represent geometrical parameters and cornering stiffness for each axis is represented by C_i. Signals of α_i are wheel side-slip angles, whose computation is $\alpha_1 = \delta - \beta - \frac{l_1\dot{\psi}}{v_x}$ and $\alpha_2 = -\beta + \frac{l_2\dot{\psi}}{v_x}$. Furthermore, the side-slip angles of the vehicle and yaw rate are β, $\dot{\psi}$, and v_y, v_x are lateral and longitudinal velocity components. Equation (8.1) can be transformed to the state-space form as

$$\dot{x}_v = A_v x_v + B_v u_v, \tag{8.2}$$

in which $x_v = \begin{bmatrix}\beta & \dot{\psi} & \psi & v_y & y\end{bmatrix}$ is the state vector, u_v contains δ and A_v, B_v represent matrices on the system. States y, ψ can be calculated through the integration of v_y, $\dot{\psi}$.

For achieving path following functionality through front wheel steering, various approaches have already been developed, e.g., in Németh et al. (2019) a method based on the Linear Parameter-Varying theory for independent steering has been proposed. The selection of the applied control method is constrained by control-oriented limitations on the test bed:

- Through the design process, the path edges among constraints to avoid lane departure, must be involved.
- Steering angle actuation has time demands, meaning low-level control and suspension movement cause delays in the system.

Accordingly, the method of Model Predictive Control (MPC) fits the objective and the constraints of the control task; see, e.g., Zhou et al. (2017) and Jalali et al. (2017). In the control process, the impact of time delay on the generation of the reference path can be compensated, which means that the formulation of the system as a time-delayed system can be avoided. The formulation of the MPC task method needs a system representation in discrete time, and, thus, the continuous-time vehicle model using a T_s sample time is reformulated. It leads to the following system description:

$$x_v(k+1) = \phi_v x_v(k) + \Gamma_v u_v(k), \tag{8.3}$$

in which representation Γ_v, ϕ_v matrices contain vehicle parameters. Vehicle motion prediction must be evaluated on the horizon n, which can be calculated as follows, see Choi and Choi (2014):

$$z_{pred}(k,n) = \begin{bmatrix} z(k+1) \\ z(k+2) \\ \vdots \\ z(k+n) \end{bmatrix} \tag{8.4}$$

$$= \begin{bmatrix} 0\,0 \\ 0\,0 \\ 1\,0 \\ 0\,0 \\ 0\,1 \end{bmatrix}^T \begin{bmatrix} \phi_v \\ \phi_v^2 \\ \vdots \\ \phi_v^n \end{bmatrix} x_v(k) + \begin{bmatrix} 0\,0 \\ 0\,0 \\ 1\,0 \\ 0\,0 \\ 0\,1 \end{bmatrix}^T \begin{bmatrix} \Gamma_v & 0 & \cdots & 0 \\ \phi_v \Gamma_v & \Gamma_v & \cdots & 0 \\ \vdots & \ddots & \ddots & \vdots \\ \phi_v^{n-1}\Gamma_v & \phi_v \Gamma_v & \cdots & \Gamma_v \end{bmatrix} \begin{bmatrix} u_v(k) \\ u_v(k+1) \\ \vdots \\ u_v(k+n-1) \end{bmatrix}.$$

Since the control objective is to ensure path following, the minimization of lateral positioning error and yaw error must be achieved. The values of the errors, i.e., y_e, ψ_e, from the vehicle states and the road reference are computed, such as

$$e(k,n) = z_{ref}(k,n) - z_{pred}(k,n), \tag{8.5}$$

in which relationship $e(k,n)$ denotes a vector, containing lateral and yaw error values. Using the computed $e(k,n)$ vector, the objective of the control, i.e., the cost function with error and control input minimization tasks, can be formulated as

$$J = \frac{1}{2} e(k,n)^T Q e(k,n) + U(k,n)^T R U(k,n), \tag{8.6}$$

where the control inputs along a horizon with n segments are contained by the vector $U(k,n) = \left[u_v(k) \ \ldots \ u_v(k+n-1)\right]^T$. Furthermore, in the cost function weighting matrices Q, R are incorporated, which ensure balancing between steering intervention and tracking performance. Using Eqs. (8.4) and (8.5), the objective is converted to

$$J = U(k,n)^T \sigma U(k,n) + \nu^T U(k,n), \tag{8.7}$$

in which function σ, ν denote matrices. This quadratic formulation is applied in the MPC problem, together with the formulation of the constraints on keeping path edges $B_b < H_{in} U < B_u$ and on the control input $l_b \leq u_i \leq l_u$. The constrained optimization problem for computing control inputs is

$$\min_{U(k,n)} U(k,n)^T \sigma U(k,n) + \nu^T U(k,n), \tag{8.8a}$$

subject to

$$B_b < H_{in} U < B_u, \tag{8.8b}$$

$$l_b \leq u_i \leq l_u, \tag{8.8c}$$

where l_b, l_u vectors ensure keeping limitation on the control input sequence. In the case of the constraint on the system states, matrix H_{in} is formed as

$$H_{in} = \begin{bmatrix} \Gamma_v & 0 & \cdots & 0 \\ \phi_v \Gamma_v & \Gamma_v & \cdots & 0 \\ \vdots & \ddots & \ddots & \vdots \\ \phi_v^{n-1}\Gamma_v & \phi_v^{n-2}\Gamma_v & \cdots & \Gamma_v \end{bmatrix}. \tag{8.9}$$

Vectors of B_b, B_u for providing constraints on the lateral motion are

$$B_b = \begin{bmatrix} x_{v,low}(t+1) \\ \vdots \\ x_{v,low}(t+n-1) \end{bmatrix}, \quad B_u = \begin{bmatrix} x_{v,up}(t+1) \\ \vdots \\ x_{v,up}(t+n-1) \end{bmatrix}, \tag{8.10}$$

in which the values of $x_{v,low}(T)$, $x_{v,up}(T)$ denote lower and upper bounds on vehicle lateral position at time step T.

The calculated vector U includes the control input sequence, which is related to the horizon, and vehicle ahead. The first element u_v in U is used in the implementation of the controller. It stands for the necessary steering angle by which the vehicle follows the road. However, u_v is not directly available, and thus, the steering angle must be produced by changing the scrub radius. Thus, from the viewpoint of the hierarchical control structure, the computed u_v is a reference, transmitted to the low-level control:

$$\delta_{ref} = u_v. \tag{8.11}$$

8.2.2 *Low-Level Control for Suspension Actuation*

The control on the low level is designed to ensure achieving δ_{ref} on the test bed, which is yielded through a linear actuator operation. Thus, the following performance criterion can be defined:

$$z_s = \delta_{ref} - \delta \qquad |z_s| \to min! \tag{8.12}$$

The basis of developing the low-level controller is to ensure the tracking performance by considering the steering dynamics of the variable-geometry vehicle suspension, described as follows:

$$\ddot{\delta} = \frac{F_l r_\delta}{I_\delta}; \tag{8.13}$$

see relationship (5.11). It is possible to treat the connection between the camber angle and the scrub radius as a linear function, such as

$$r_\delta = k_1 \gamma, \tag{8.14}$$

where k_1 is a constant parameter. Furthermore, the connection of actuator position and wheel tilting angle is approximated through a relationship in linear form (Németh et al. 2019):

$$\gamma = k_2 d, \tag{8.15}$$

where k_2 is a parameter and d is the linear actuator position.

Accordingly, the dynamics related to the relation between the angle of steering and actuator position is described by

$$\ddot{\delta} = \frac{F_l k_1 k_2 d}{I_\delta}. \tag{8.16}$$

For the goals of control design, it is possible to convert the introduced dynamics into Laplace domain (s):

$$G(s) = \frac{\Delta(s)}{D(s)} = \frac{F_l k_1 k_2}{I_\delta s^2} = \frac{K}{s^2}, \tag{8.17}$$

where $D(s)$ and $\Delta(s)$ signals result from the Laplace transformation of $d(t)$, $\delta(t)$. Consequently, steering dynamics contains a double integration due to $\frac{1}{s^2}$ formula. Thus, a reasonable selection can be a derivative controller that provides tracking of the reference signal as well as speeding up the actuation of the controller. The $C(s)$ controller is described as follows:

$$C(s) = Ps, \tag{8.18}$$

in which control rule P is selected as a tuning parameter; see Fig. 8.3. The proposed $C(s)$ controller has one input e_δ, which is the error in the tracking of steering angle, and its output is linear actuator position command d.

The proposed $C(s)$ controller is implemented in discrete time for the operation of the HiL environment, and thus, the controller (8.18) is realized with a sampling time of $T_{s,2}$. Nevertheless, during the realization process its output must be transformed, because two interventions of the electronic board of the linear actuator are available (see Sect. 8.1.1), i.e., maximum speed intervention in both directions. Due to these constraints on the intervention, a cascade architecture for the low-level control is designed: in the outer loop a P controller is included, while the role of inner loop C_{in} is

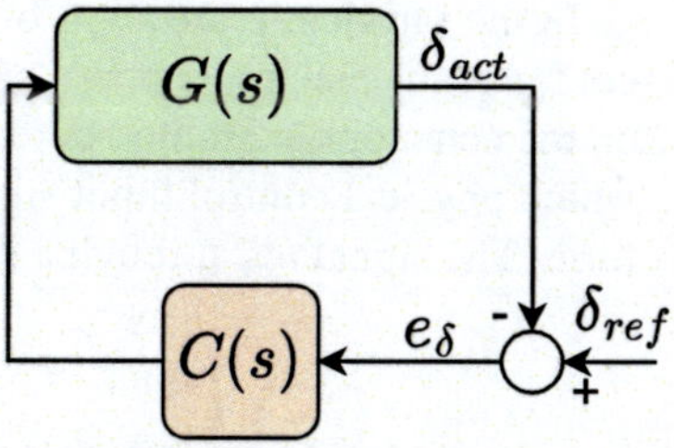

Fig. 8.3 Illustration of the outer loop for low-level controller

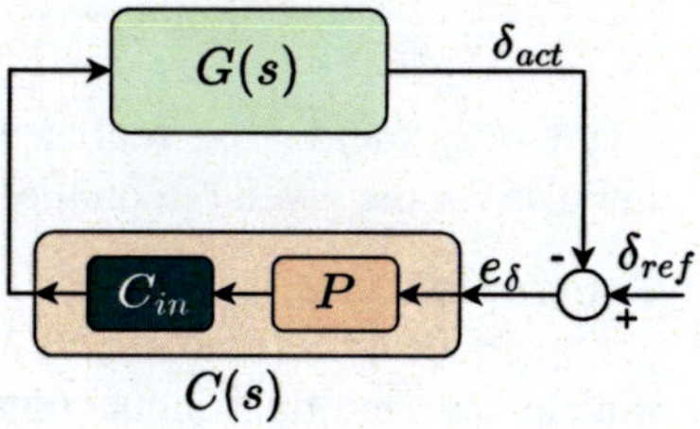

Fig. 8.4 Scheme on cascade architecture for control on the low level

to handle actuator limits for avoiding the effect of chattering. Figure 8.4 illustrates the proposed control architecture for low-level operation. In the presented solution, the inner loop controller C_{in} consists of two main parts, such as a numerical integration and a numerical differentiation with a dead-zone, as follows. It is defined as a signal d_{ref}, which is a reference for tracking of actuator position. Its value can be calculated through the numerical integration e_δ, i.e., summation of inner loop input:

$$d_{ref}(k) = P\sum_{j=1}^{k}(e_\delta(j))T_{s,1} = P\sum_{j=1}^{k}(\delta_{ref}(j) - \delta(j))T_{s,1}, \qquad (8.19)$$

where P denotes the tuning parameter in (8.18) and T_{s_1} represents sample time in the inner loop control. The computed d_{ref} signal is used for formulating the input of the dead-zone:

$$u_{dz} = (d_{ref} - d)/T_{s,1}. \qquad (8.20)$$

Remark 8.1 In the case of $P = 1$ selection with very small $T_{s,1}$ time value, expression (8.19) is converted to an integral expression: $\int_0^t e_\delta(\tau)d\tau$. Because fast position tracking can be considered, the expression of the integral is $\int_0^t e_\delta(\tau)d\tau = d(t)$, which results in $e_\delta(t) = \dot{d}(t)$, similar to (8.16). Thus, this algorithm can be considered a derivative controller with respect to e_δ. Sample time values for inner and outer loops must be selected as $T_{s,1} \ll T_{s,2}$, with which the assumption of fast positioning can be yielded.

In the physical realization of the control system, the calculated signals u_{dz} are not fed to the electronic board of the linear actuator. There are three designated pins on the actuator electronic board for actuator control, such as stop, down and up, in which physical control input signals are arranged in $U_s = [up, down, stop]$ vector form. The signal u_{dz} through a dead-zone to impulse-formed signals is converted:

$$\begin{cases} \text{if } u_{dz}(n) > dz_{high} \text{ then } U_s = [1, 0, 0], \\ \text{if } u_{dz}(n) < dz_{low} \text{ then } U_s = [0, 1, 0], \\ \text{otherwise, } \quad U_s = [0, 0, 1], \end{cases} \tag{8.21}$$

where dz_{high} and dz_{low} represent dead-zone boundaries, and notation 1 represents impulse for the given pin on the electrical board.

Remark 8.2 It is illustrated in the dynamic equation that the longitudinal force acting on the wheel also affects δ. Accordingly, the steering angle setting can be formulated as a multiple-input control task, which requires the coordination of camber angle modification and driving torque. Although its coordination method is proposed in Chap. 7, due to limitations of the test bed on high-precision driving torque it has been eliminated. Thus, in the case of the implementation only camber angle modification is realized.

8.3 Illustration of Tuning Parameter Selection

The presented test bed includes a number of nonlinearities and uncertainties that makes it difficult to tune the low-level controller using a systematic method. Therefore, parameter P is chosen through test scenarios, experimentally (Fényes et al. 2022). Figure 8.5 illustrates the relative error of $\dot{d}$ tracking in a test scenario, in which a sinusoidal reference and $P = 1$ are selected. It is carried out with 300 RPI wheel rotational speed, which leads to $v_x = 25$ km/h longitudinal velocity in the HiL simulation. It can be seen that the internal loop control algorithm can ensure the desired fast tracking performance, due to the limited relative error on the tracking of $\dot{d}$.

Using the results of the preliminary test scenarios, the value of P is selected. During the tests P can vary in the range of $P \in \{30 - 100\}$. Illustrative examples of the achieved steering angle tracking performance can be seen in Fig. 8.6. As is shown, smooth tracking with low P value can be achieved, though low-value selection for P can lead to increased time delay. Numerical results on the analysis of time delays and tracking errors with various P values are as follows. For $P = 30; 50; 70; 100$ selections each average positioning error values are 1.0072°; 1.0209°; 1.2028°; 2.0442° and the average delay values are 1.11 s; 0.75 s; 0.68 s; 0.61 s. As a result, the smallest averaged error is for the case of $P = 30$, and $P = 100$ leads to significant steering angle tracking error, although the trend of the achieved time delay is reversed, i.e.,

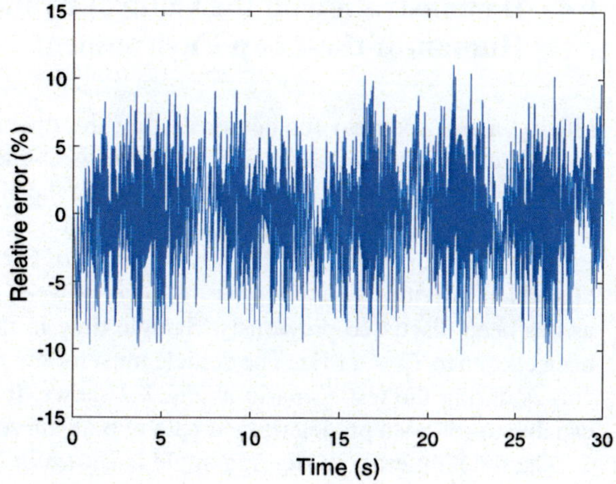

Fig. 8.5 Illustration of a test scenario for inner loop control testing

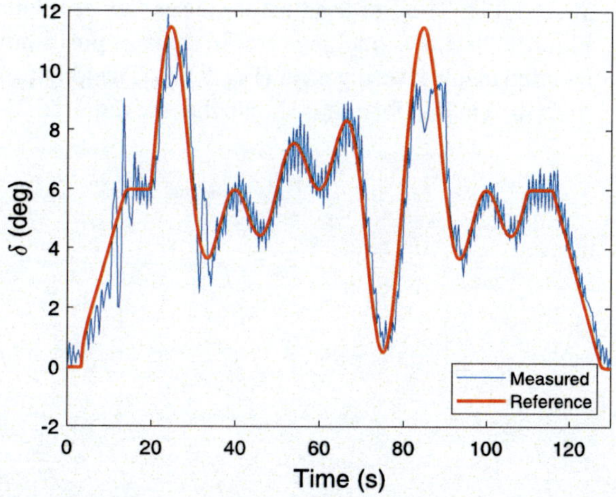

Fig. 8.6 Illustration of the test results with $P = 40$ selection

$P = 30$ leads to increased delay and $P = 100$ results in low delay. Thus, it is necessary to find a balance in the simultaneous reductions of time delay and tracking error.

It can be seen that even the shortest delay is notable ($t_d = 0.61$ s), which has resulted in physical limitations on the actuator control, i.e., motion properties. Therefore, the value of delay in tracking has a minimum, which cannot be further reduced. A possible solution for the compensation of t_d is to shift forward the reference signal on the path, i.e., the problem may be handled on the high-level control.

8.4 Demonstration on the Control Evaluation Under Human-in-the-Loop Environment

Finally, a test scenario for demonstrating the operation of the variable-geometry suspension control in this section is shown. A D-class passenger vehicle weighing 1, 534 kg has been selected for simulation. In the HiL simulation, one of its front wheels is considered to be connected to the test bed, i.e., measurements and control interventions for one wheel have been realized. Parameter P has been selected to 50, which provides an acceptable balance between time delay and tracking error, as has been discussed previously. Sample time in the operation of the control has been chosen to $T_s = 1$ kHz. The vehicle must follow a predetermined path with sharp curves during the test scenario as Fig. 8.7 shows. It can be seen that the vehicle is capable to follow a predetermined path at both curves, with low lateral error.

The tracking error in steering angle is shown in Fig. 8.8a. Tracking the steering angle is precise, but there are some oscillations in the straight sections, e.g., between 30 and 55 s. The linear actuator causes this oscillation as it operates near its limits. Figure 8.8b shows tracking error in actuator positioning has a very small value, which resulted in low steering error (Fig. 8.8a). The longitudinal speed error of the vehicle is described in Fig. 8.8c. It is shown that the required 25 km/h longitudinal speed during

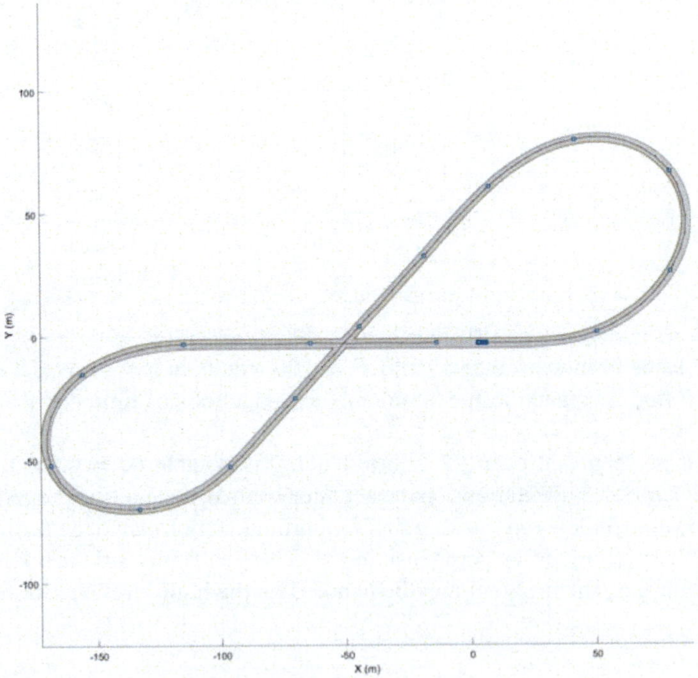

Fig. 8.7 Illustration of road and vehicle path

Fig. 8.8 Illustration of tracking errors

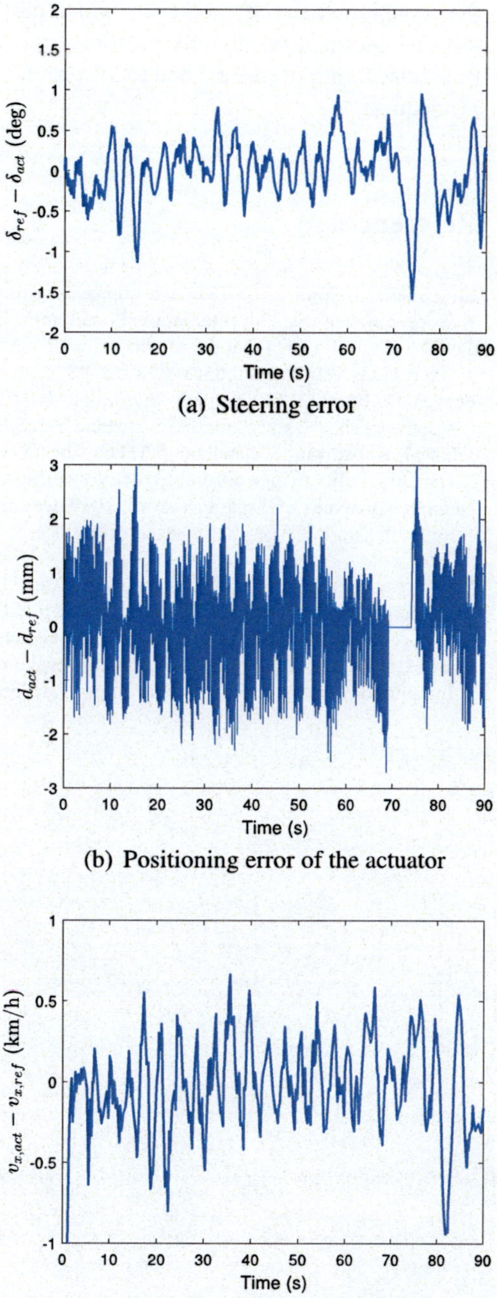

(a) Steering error

(b) Positioning error of the actuator

(c) Speed error of the vehicle

the simulation has been achieved, which corresponds to the limited capability of the test bed. A small oscillation is observed in the measurement that is caused by the imprecise fixing of the test bed to the ground, but the achieved tracking performance is adequate.

References

Chamraz S, Balogh R (2018) Two approaches to the adaptive cruise control (acc) design. In: Proceedings of the 29th international conference 2018 cybernetics and informatics, pp 1–6

Choi M, Choi SB (2014) Model predictive control for vehicle yaw stability with practical concerns. IEEE Trans Veh Technol 63(8):3539–3548

Fényes D, Fazekas M, Németh B, Gáspár P (2022) Implementation of a variable-geometry suspension-based steering control system. Veh Syst Dyn 60(6):2018–2035

Jalali M, Khosravani S, Khajepour A, ken Chen S, Litkouhi B (2017) Model predictive control of vehicle stability using coordinated activesteering and differential brakes. Mechatronics 48:30–41

Németh B, Fényes D, Gáspár P, Bokor J (2019) Coordination of independent steering and torque vectoring in a variable-geometry suspension system. IEEE Trans Control Syst Technol 27(5):2209–2220

Sahputro SD, Fadilah F, Wicaksono NA, Yusivar F (2017) Design and implementation of adaptive pid controller for speed control of dc motor. In: 2017 15th international conference on quality in research (QiR): international symposium on electrical and computer engineering, pp 179–183

Zhou H, Guvenc L, Liu Z (2017) Design and evaluation of path following controller based on MPC for autonomous vehicle. In: Proceedings of the 36th Chinese control conference, pp 9934–9939

Part III
Guaranteed Suspension Control with Learning Methods

Chapter 9
Data-Driven Framework for Variable-Geometry Suspension Control

This chapter has been motivated the providing of high-precision models for low-level dynamics of variable-geometry vehicle suspensions. The precise positioning of the wheel, such as performing a steering angle, requires precise control-oriented modeling of the test equipment that can be utilized for design purposes. Nevertheless, through the implementation of the controller there are some specialties of the physical system, which are not possible to involve in the model; e.g., determining high-frequency physical disturbances of the equipment is very difficult, as it depends on the speed of the wheel, the actual loads and the flexibility of the structure. Furthermore, unmodeled dynamics may occur in the suspension joints and the friction between the rotating plate and the tire.

A possible solution on the modeling of the test bed is to define nonlinear effects, which can help solve the problem by high-complexity physical-based equations. However, this process can have some drawbacks. One is that the complex mechanical formulas need many precise parameters, which can cause difficult parameter identification problems. The other is that it can be problematic designing controllers with several nonlinear relations in the system. To avoid these issues, a data-based method for modeling the presented physical suspension equipment is introduced. Earlier, data-driven methods have been successfully applied in the context of suspension systems, as described in Formentin and Karimi (2013), Radac and Lala (2020) and Freddy and Fredy (2019).

This chapter focuses on the modeling procedure of steering, based on a data-driven method. Moreover, the resulting control-oriented state-space model is used for LPV-based control design. During modeling, a pace regression is used with a subset selection algorithm to discover the relationships among the measured signals on the system. As a result, a series of linear systems are formed that are applied for creating the LPV structured model. In the control design, the resulting control-oriented model is involved, with which low-level performance requirements are fulfilled, i.e., to

B. Németh and P. Gáspár, *Control of Variable-Geometry Vehicle Suspensions*, Advances in Industrial Control, https://doi.org/10.1007/978-3-031-30537-5_9

follow a predetermined reference steering angle and to reduce the movement of the linear actuator, which is the control signal of the system.

The suggested solution has the benefit to avoid the difficulties of complex modeling, and therefore control-oriented modeling can be achieved. Furthermore, the proposed technique may be utilized to solve additional model formulation or control design issues, i.e., it provides the generalization possibility. Nevertheless, the method has the disadvantage of needing a large number of measured datasets, and, simultaneously, the resulting control-oriented model may not be related to the physical-based model, e.g., in their parameters. Moreover, modifying the structure of the physical equipment (for example, replacing a constructional unit) may need training of the agent.

9.1 Control-Oriented Model Formulation of the Test Bed

This section briefly presents the method of pace regression, as a basic approach to data-driven modeling and control. It is considered a dataset with n independent instances, input variables with k number and an output variable. In the dataset, the instances in an $n \times k$ matrix, which is noted with X, are arranged. Furthermore, the set of parameters of the real model is ordered on a vector ζ^*. The output vector y of the real system is defined as follows:

$$y = X\zeta^* + \varepsilon, \tag{9.1}$$

in which ε represents the vector of noises, whose elements are sampled as $N(0, \sigma^2)$, where σ^2 is considered to be known or estimated ($\hat{\sigma}^2$). The compact description of the process through a linear model is denoted by $\mathscr{M}(\zeta)$ with ζ unique vector of the approximated parameters, and, similarly, the process model with the real parameters is represented by $\mathscr{M}(\zeta^*)$. The model fitting task has the goal to find a model $\mathbb{M} = \{\mathscr{M}(\zeta) : \zeta \in \mathbb{R}^k\}$, which leads to the lowest error on the prediction of the process, regarding the given measured dataset. Several methods can be used for generating models, e.g., Ordinary Least Square (OLS) method, OLS subset selection, shrinkage, RIC and CIC methods, as detailed in Wang and Witten (1999). These techniques can decrease the dimension of models by discarding redundant variables. To evaluate the model, it is necessary to know the distance between the current and the real model. The distance can be computed as follows:

$$\mathscr{D}(\mathscr{M}(\zeta^*), \mathscr{M}(\zeta)) = \frac{||y_{\mathscr{M}(\zeta^*)} - y_{\mathscr{M}(\zeta)}||^2}{\sigma^2}, \tag{9.2}$$

in which $\mathscr{L}_2$ norm is noted with $|| \cdot ||$. Moreover, the value of σ^2 is replaced by the estimated value $\hat{\sigma}^2$. Thus, the model fitting task is formed as the following minimization problem (Fényes et al. 2021):

$$\mathscr{D}(\mathscr{M}^*, \mathscr{M}) = min! \tag{9.3}$$

As a fundamental idea behind OLS-based selection method of subsets, generation of subset models by different variable sets is requested. For example, a dataset with k number of variables, the requested number of nested models ($\mathscr{M}_j$) is $k+1$. In notation $\mathscr{M}_j$, the null model with zero variables results in $j=0$, and in the case of $j=k$ all the variables are used, and, thus, the parameter vector for $\mathscr{M}_j$ can be estimated as

$$\hat{\xi}_{\mathscr{M}_j} = (X'_{\mathscr{M}j} X_{\mathscr{M}j})^{-1} X'_{\mathscr{M}j} y, \tag{9.4}$$

where $X_{\mathscr{M}_j}$ has the size $n \times j$. Moreover, $\mathscr{P}_{\mathscr{M}_j}$ is an orthogonal projection matrix, which creates a connection between the original space (k) and the decreased space (j), such as $\mathscr{P}_{\mathscr{M}_j} = X_{\mathscr{M}j}(X'_{\mathscr{M}j} X_{\mathscr{M}j})^{-1} X'_{\mathscr{M}j}$. Consequently, $\hat{y}_{\mathscr{M}_j} = \mathscr{P}_{\mathscr{M}_j} y$ estimation can be formulated.

For providing a full solution, 2^k number of subset models must be generated. Nevertheless, in case of increasing k values, its computation is unfeasible. To avoid this problem, the presented procedure generates $k+1$ subset models, which must be ranked to find the best model. Details on the ranking algorithms for subset selection can be found in Shibata (1981) and Thompson (1978).

Figure 9.1 shows an exemplary real measurement, which is used for the model fitting task. It is demonstrated that steering angle $\delta \in \{-6°, 0°\}$ is achieved with actuator movement $d \in \{0 - 300\}$ mm, and the steering angle between $\delta \in \{0°, 6°\}$ with motion $d \in \{300 - 500\}$ mm can be reached. The interconnection of d, δ implies the characteristics with nonlinearities. Thus, the control-oriented linear model formulation of the entire system can require its division into subsystems, in function of the steering angle value. Thus, it is necessary to find models in the linear form, i.e., through the pace regression technique. Moreover, through their interconnection a LPV model can be created.

The proposed data-driven model formulation requests a large number of test scenarios, whose measurements must be ordered into a structured dataset. Thus, for creating adequate models, many test cases with varying wheel speed have been carried out within the range 300–900 RPM. Simultaneously, the actuator moved randomly up and down during the tests. Accordingly, approximately 500,000 individual instances through the test have been logged, within the signals: steering angle δ, position of the actuator d and speed of the wheel $v_x(t)$. Model formulation through the following process has been carried out:

- The dataset has been separated into various subsets, depending on the actual steering angle. In this example, the range of steering has been divided with the steps of 2°.
- In the next step, the dataset has been processed for model formulation. The deviations of δ and d have been calculated for each measured instance in the dataset:

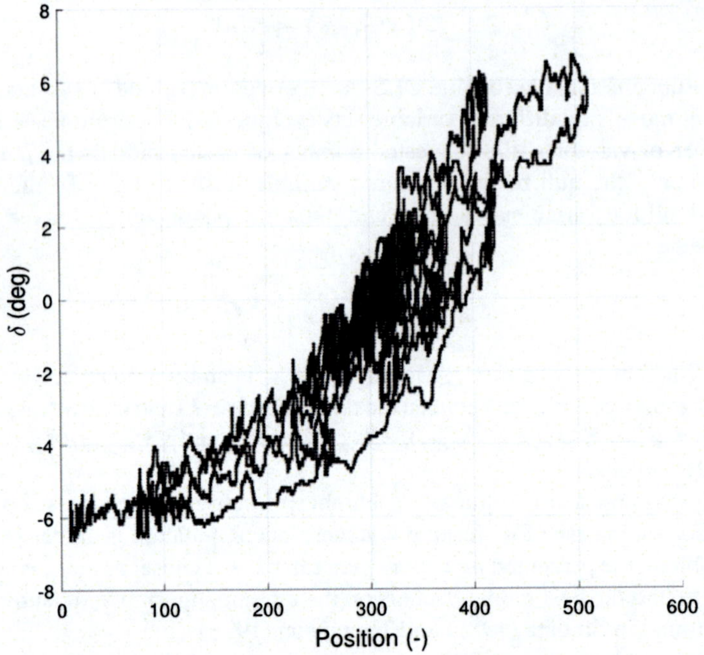

Fig. 9.1 Exemplary measurement to illustrate the connection of actuator position and δ

$$\Delta\delta = \delta(t) - \delta(t-1), \tag{9.5a}$$

$$\Delta d = d(t) - d(t-1). \tag{9.5b}$$

The reformulation is motivated to achieve increased correlation, i.e., the relation between $\Delta\delta$ and Δd can be easily formulated. This is caused by the initial values of the test bed settings, which influence the absolute values of δ and d significantly.

- Third, connection between current $\Delta\delta(t)$ and the system attributes is established. The connection consists of $\Delta d(t)$, $v_x(t)$ and, moreover, v_x, $\Delta\delta$ and Δd with their past values. The model form for calculating $\Delta\delta$ has been provided as

$$\hat{y}_{\mathcal{M}_j} = X_{\mathcal{M}_j}\hat{\zeta}_{\mathcal{M}_j}, \tag{9.6}$$

where $\hat{y}_{\mathcal{M}_j}$ is equal to $\Delta\delta(t)$. Vector $\hat{\zeta}_{\mathcal{M}_j}$ consists of $\Delta\delta(t-1)\ldots\Delta\delta(t-n)$, $\Delta d(t)\ldots\Delta(t-n)$, $v_x(t)\ldots v_x(t-n)$, with $n=5$. Moreover, $X_{\mathcal{M}_j}$ includes the real test bed parameter values. The model (9.6) has been created for all of the subsets related to δ.

- Finally, the coefficients in X have been selected for all subsets through the solution of various optimization tasks. Finding the best model of each subset is computed as

$$\min_{X} (\hat{y} - y)^2, \tag{9.7}$$

in which optimization problem y is current $\Delta\delta(t)$ measured signals on the test bed. This optimization task is performed using the method of the proposed pace regression, joining with the subset selection method; see Wang and Witten (1999).

The resulting models in (9.6) form are validated through the k-fold method. For every subset at the selected steering angle values, through pace regression technique linear models are generated, whose correlation coefficient is above >0.985 in the given example. The set of linear models describe the full dynamics of the test bed. As the last step of the model formulation, the set of systems is described by polytopic LPV approach. The state-space model in a discrete form is

$$x_s(t+1) = A_s(\rho)x_s(t) + B_{s1}(\rho)u_s(t) + B_{s2}(\rho)\omega_s(t), \tag{9.8}$$

in which x_s vector contains the elements of $\hat{y}_{\mathcal{M}_j}$ and δ, while $u_s = \Delta d$. State $\delta(t)$ in x_s is calculated through the expression $\delta(t) = \sum_{T=0}^{t} \Delta\delta(T)$. Furthermore, $\omega_s = v_x$ is the wheel speed that is considered as a disturbance of the system. Moreover, the ρ scheduling variable is equal to the averaged δ value, around which the linear models of each subset are fitted.

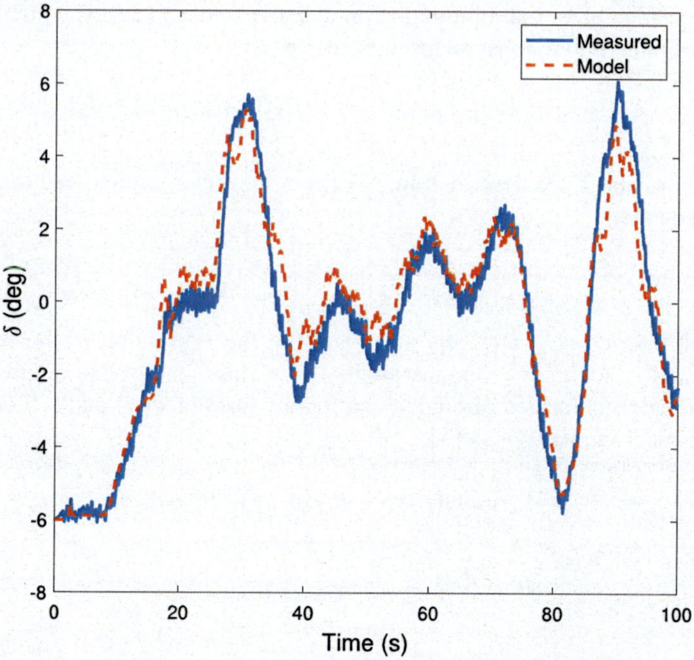

Fig. 9.2 Example of the comparison of the measured and modeled steering angles

In Fig. 9.2, the model fitting precision based on a test scenario is demonstrated, i.e., measured and modeled steering angles are shown. It is well demonstrated that steering angles are close to each other, i.e., the error between the measured δ and the fitted δ with the data-driven method is small, which leads to an error under 0.24°. Consequently, a high-precision model of the test bed dynamics is achieved (Fényes et al. 2021).

9.2 Design of LPV Control to Achieve Low-Level Operations

On the level of the suspension actuator, the linear reference steering angle δ_{ref} must be achieved to guarantee the motion of the vehicle. In the design, the resulting LPV model (9.8) is used in the control design process applied (Gáspár et al. 2017). Through the control design, the next performance criteria have to be ensured:

- Tracking of δ_{ref} is the most important performance criteria against the suspension. Value of reference angle has been generated by the controller on the upper level. These criteria lead to the minimization of the following error signal:

$$z_1 = \delta_{ref} - \delta, \qquad |z_1| \rightarrow min. \tag{9.9}$$

- The linear actuator has a limitation on its movement capability, which means that control intervention must be minimized, i.e.,

$$z_2 = d, \qquad |z_2| \rightarrow min. \tag{9.10}$$

The performances are described in $z = \begin{bmatrix} z_1 & z_2 \end{bmatrix}^T$ vector, which results in a performance equation:

$$z = C_1 x_s + D_{11} r_s + D_{12} u_s, \tag{9.11}$$

in which matrices C_1, D_{11}, D_{12} are related to the expression of the performance vector, and r_s includes reference angle δ_{ref}. Consequently, the augmented state-space equation is formed through a continuous form of (9.8) and (10.18), together with the measurement $y_K = [\delta]$:

$$\dot{x}_s = A_s(\rho)x_s + B_{s1}(\rho)u_s + B_{s2}(\rho)\omega_s, \tag{9.12a}$$

$$z = C_1 x_s + D_{11} r_s + D_{12} u_s, \tag{9.12b}$$

$$y_K = C_2 x_s. \tag{9.12c}$$

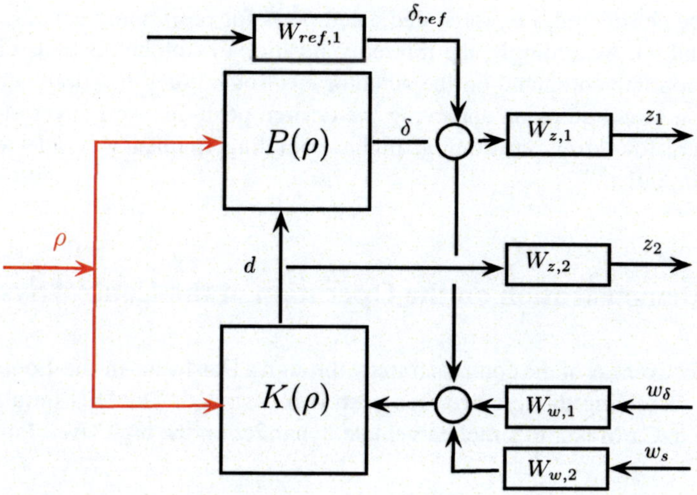

Fig. 9.3 Illustration of the augmented plant

During the design of the control system, weighting functions on (9.12) are used to achieve the requested performance level and to scale disturbances. The purpose of $W_{z,1}$ is to ensure the exact reference tracking capability. $W_{ref,1}$ is used for scaling δ_{ref}, which is achieved from the controller on the upper level. For scaling the intervention $W_{z,2}$ is applied, while $W_{w,1}$ has been used to reject noises on the steering angle sensor. Finally, weight $W_{w,2}$ has been applied to compensate for the impact of v_x on steering angle measurement. The structure is illustrated in Fig. 9.3.

The performance-driven LPV-based control design task is to choose parameter-varying controller $K(\rho)$ to guarantee the quadratic stability of the closed-loop system, and to ensure $\mathscr{L}_2$ norm from $w = [\omega_s,\ w_\delta,\ w_s]^T$ to performances under the constant value γ. The control design problem is given as follows (Wu 2001):

$$\inf_{K(\rho)} \sup_{\rho \in F_\rho} \sup_{\|w\|_2 \neq 0, w \in \mathscr{L}_2} \frac{\|z\|_2}{\|w\|_2}, \tag{9.13}$$

where F_ρ limits the scheduling variables (Packard and Balas 1997). The resulting $K(\rho)$ controller is configured as follows:

$$\dot{x}_K = A_K(\rho)x_K + B_K(\rho)y_K, \tag{9.14a}$$

$$u = C_K(\rho)x_K + D_K(\rho)y_K, \tag{9.14b}$$

where the variable-dependent matrices are $A_K(\rho)$, $B_K(\rho)$, $C_K(\rho)$, $D_K(\rho)$. When implementing a controller in dSpace Autobox, the resulting control system has to be transformed into a discrete-time system (Tóth 2010). Sampling time in the case of the test equipment has been set to $T_{s,1} = 0.01$ s. Furthermore, during the control implementation the input characteristics of the electronic board on the linear actuator has to be considered, i.e., three dedicated ports for positioning exist, such as stop, down and up. Accordingly, the reference position to Autobox is transmitted. Then, the appropriate command on the actuator electronic board has been activated, and the linear motor stops on achieving the desired position; see Fényes et al. (2022). To ensure low error positioning, higher sampling frequency is selected, such as $T_{s,2} = 0.0001$ s.

9.3 Demonstration on the Operation of the Control System

The effectiveness of the control strategy through a Hardware-in-the-Loop (HiL) simulation, involving the test bed, has been demonstrated. The HiL simulation incorporates in CarMaker to simulate vehicle dynamics on the high level. For simulation

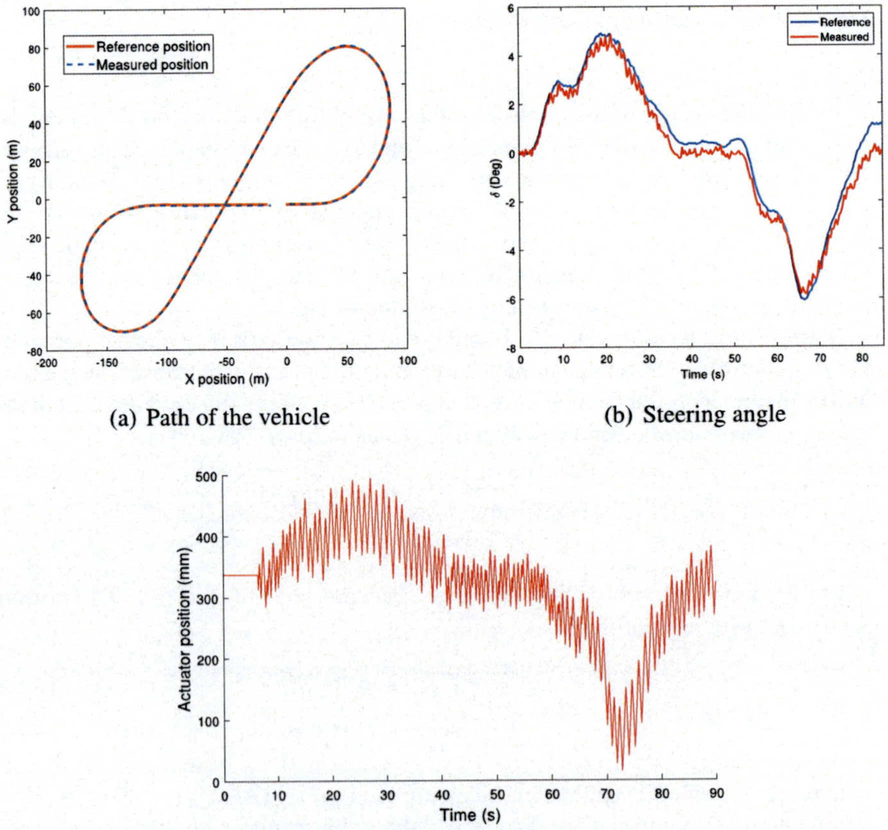

(a) Path of the vehicle

(b) Steering angle

(c) Position of the actuator

Fig. 9.4 HiL simulation results

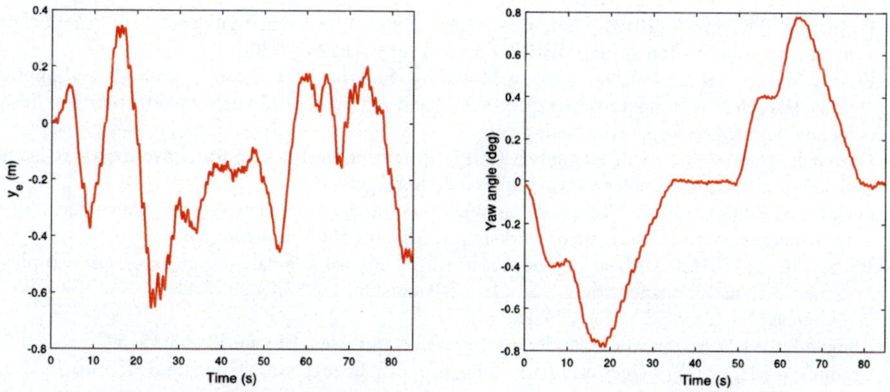

(a) Error of lateral position

(b) Error of yaw-angle

Fig. 9.5 Tracking errors

purposes, a medium-sized 1538 kg passenger vehicle is selected. Figure 9.4a shows the motion of the vehicle with constant speed, i.e., on a path consisting of two curves (one to the left and one to the right).

It can be seen that the introduced control system is capable of guaranteeing appropriate path following. Signal δ_{ref} together with the measured steering angle is illustrated in Fig. 9.4b, which is calculated on the high level of the control hierarchy. The maximum tracking error is less than <0.4 deg, and thus, the error is small. This demonstrates that steering functionality on the level of the test bed can be ensured with the proposed control technique based on a data-driven model. The position of the actuator, i.e., the motion of the linear motor is illustrated in Fig. 9.4c.

Further results on the operation of the implemented control are shown in Fig. 9.5. Figure 9.5a shows the lateral error, whose maximum is 0.6 m. This value is caused by the physical limitations (e.g., position and velocity) of the linear motor, which are involved in the control design on the high level; see Fényes et al. (2022). Figure 9.4c shows that the actuator has also a limitation on its velocity, which leads to limited oscillation in the actuator position signal. The yaw-rate error of the vehicle, which is related to the geometry of the centerline, has a limited small value (<0.8 deg); see Fig. 9.5b.

References

Fényes D, Németh B, Gáspár P (2021) Data-driven modeling and control design in a hierarchical structure for a variable-geometry suspension test bed. In: 2021 60th IEEE conference on decision and control (CDC), pp 5047–5052

Fényes D, Fazekas M, Németh B, Gáspár P (2022) Implementation of a variable-geometry suspension-based steering control system. Veh Syst Dyn 60(6):2018–2035

Formentin S, Karimi A (2013) A data-driven approach to mixed-sensitivity control with application to an active suspension system. IEEE Trans Ind Inf 9(4):2293–2300

Freddy V, Fredy R (2019) Data-Driven Modeling, Filtering and Control: Methods and applications, IET, chap A comparative study of VRFT and set-membership data-driven controller design techniques: active suspension tuning case

Gáspár P, Szabó Z, Bokor J, Németh B (2017) Robust control design for active driver assistance systems: a linear-parameter-varying approach. Springer

Packard A, Balas G (1997) Theory and application of linear parameter varying control techniques. In: American control conference, workshop I, Albuquerque, New Mexico

Radac MB, Lala T (2020) Learning nonlinear robust control as a data-driven zero-sum two-player game for an active suspension system. IFAC-PapersOnLine 53(2):8057–8062, 21st IFAC World Congress

Shibata R (1981) An optimal selection of regression variables. Biometrika 68:45–54

Thompson ML (1978) Selection of variables in multiple regression. Int Stat Soc B 46:1–21 and 129–146

Tóth R (2010) Modeling and identification of linear parameter-varying systems, vol 403. Lecture notes in control and information sciences. Springer, Berlin, Heidelberg

Wang Y, Witten IH (1999) Pace regression. (Working paper 99/12). Hamilton, New Zealand: University of Waikato, Department of Computer Science

Wu F (2001) A generalized lpv system analysis and control synthesis framework. Int J Control 74:745–759

Chapter 10
Guaranteeing Performance Requirements for Suspensions via Robust LPV Framework

In this chapter, a control design structure for variable-geometry suspension systems is proposed, in which structure non-standard control elements, such as neural networks, can be included. The essential of this structure is that the minimum performance level of selected performances can be guaranteed, even at the undesired operation of the unconventional element. Because of the motivation of automated vehicle systems, the unconventional element in this book is the machine learning-based controller.

10.1 Fundamentals of the Control Design Structure

In the proposed control strategy, two control elements are considered. First, it is given a non-conventional control, e.g., a neural network, which can be formed as follows:

$$u_L = \mathscr{F}(y_L), \tag{10.1}$$

in which relationship $\mathscr{F}$ denotes the learning-based control. The controller has m_L number of inputs in the vector y_L, and its output, i.e., control input from the learning-based control, is $u_L = \left[u_{L,1}\ u_{L,2}\ \ldots\ u_{L,n}\right]^T$ with n outputs. Second, in this structure a model-based control is involved:

$$u_K = \mathscr{K}(\rho_K, y_K), \tag{10.2}$$

where $\mathscr{K}$ denotes a robust Linear Parameter-Varying (LPV) control, which has scheduling vector ρ_K with f number of variables, i.e., $\rho_K \in \Omega_K$ bound is defined. For the robust LPV control, the vector $u_K = \left[u_{K,1}\ u_{K,2}\ \ldots\ u_{K,n}\right]^T$ represents its output, which is another control input vector of the system. Moreover, y_K vector involves the m_K number of measurements on the system.

B. Németh and P. Gáspár, *Control of Variable-Geometry Vehicle Suspensions*, Advances in Industrial Control, https://doi.org/10.1007/978-3-031-30537-5_10

The design process basically assumes that $u = \left[u_1 \dots u_n\right]^T$ input vector for the system is generated by u_K and u_L. It has been achieved through the definition of u in a special form, i.e., it is given in a linear formula, depending on u_K and further predetermined terms. The linear formulation parameters are chosen to ensure that u is inside of a predetermined region of u_K, and, similarly, u is as close as possible to u_L. Considering that $u = u_L$ can be achieved, the linear form is defined as

$$u = I_{n \times n} \circ (\rho_L^* J_{1 \times n}) u_K + \Delta_L^*, \tag{10.3}$$

in which definition Hadamard product is represented by $\circ$, $J_{1 \times n}$ denotes a one-row vector and $I_{n \times n}$ denotes $n \times n$ identity matrix. Moreover, Δ_L^* and ρ_L^* vectors with n terms are formed as

$$\Delta_L^* = \left[\Delta_{L,1}^* \; \Delta_{L,2}^* \dots \Delta_{L,n}^*\right]^T, \quad \Delta_{L,i}^* \in \Lambda_{L,i}, \tag{10.4a}$$

$$\rho_L^* = \left[\rho_{L,1}^* \; \rho_{L,2}^* \dots \rho_{L,n}^*\right]^T, \quad \rho_{L,i}^* \in \Omega_{L,i}, \tag{10.4b}$$

where $\Delta_{L,i}^*$ and $\rho_{L,i}^*$ are time-varying signals with the bounds $\Lambda_{L,i} = [\Delta_{L,i,min}; \Delta_{L,i,max}]$, $\Omega_{L,i} = [\rho_{L,i,min}; \rho_{L,i,max}]$ on each signal, involved in the vectors Λ_L, Ω_L.

In case of guaranteeing all conditions in (10.3), the equality of u and u_L can be achieved. Otherwise, $u \neq u_L$ because $\Delta_{L,i}^*$, $\rho_{L,i}^*$ have limits due to Λ_L, Ω_L boundaries. Therefore, in the general case the form of the control input is (Németh and Gáspár 2021b)

$$u = I_{n \times n} \circ (\rho_L J_{1 \times n}) u_K + \Delta_L, \tag{10.5}$$

in which formalization

$$\rho_L = \left[\rho_{L,1} \; \rho_{L,2} \dots \rho_{L,n}\right]^T, \tag{10.6a}$$

$$\Delta_L = \left[\Delta_{L,1} \; \Delta_{L,2} \dots \Delta_{L,n}\right]^T, \tag{10.6b}$$

$$\rho_{L,i} = \min\left(\max\left(\rho_{L,i}^*; \rho_{L,i,max}\right); \rho_{L,i,min}\right), \quad \forall i = 1 \dots n, \tag{10.6c}$$

$$\Delta_{L,i} = \min\left(\max\left(\Delta_{L,i}^*; \Delta_{L,i,min}\right); \Delta_{L,i,max}\right), \quad \forall i = 1 \dots n. \tag{10.6d}$$

Formulations (10.6c)–(10.6d) ensure containing ρ_L, Δ_L in their related domains.

Figure 10.1 shows the connections of u, u_L and u_K within the architecture of the control. The LPV control in the feedback loop is incorporated, but the non-conventional control system in another auxiliary loop is involved. Selecting ρ_L, Δ_L is the purpose of the selection block in the presented structure.

The contribution of the presented control is the guaranteeing of the minimum level of performance, which through the LPV control over the whole system operating range is determined. By guaranteeing all conditions in (10.3), the level of control

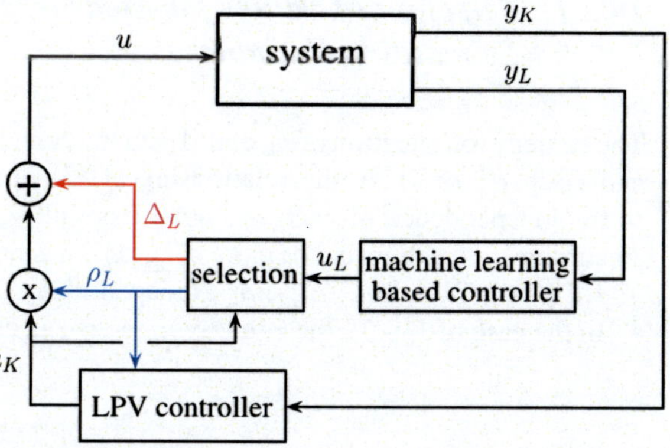

Fig. 10.1 Illustration of the architecture of the control

performances through the learning-based control is determined, i.e., improved. Thus, using this control strategy, the minimum control performance level is guaranteed, and, similarly, through the non-conventional control its maximum level can be improved.

The following design steps for achieving the control architecture must be carried out:

1. The non-conventional control through a learning-based approach must be designed, which yields $\mathscr{F}$ for variable-geometry suspension.
2. Boundaries of domains Ω_L, Λ_L must be selected, in which the operation with guarantees can be achieved.
3. Design of suspension controller $\mathscr{K}$ for steering purposes using the robust LPV theory is carried out, considering the domains of Λ_L, Ω_L in the synthesis process.

A difficulty of the proposed design process is the relationship between the LPV control design and the selection of the bounds Ω_L, Λ_L; i.e., it is necessary to select the domains during control synthesis, while efficient domain boundary selection needs knowledge on the performance of $\mathscr{K}$. A solution for handling this relationship is to design the LPV control and to select Λ_L, Ω_L using an iterative process.

10.2 Selection Process for Measured Disturbances and Scheduling Variables

In the case of measured disturbances and scheduling variables, two processes for efficient selection must be performed. It is necessary to select ρ_L, Δ continuously along control system operation (i.e., in real time) for creating u from u_L, u_K. Moreover, domains of Λ_L, Ω_L during the iterative process (i.e., offline) must also be selected.

10.2.1 Selection of Values for Measured Disturbances and Scheduling Variables

The strategy for selection of ρ_L and Δ_L on the connection of u_K, u_L is based (Németh and Gáspár 2021a). In the relationship (10.3), the term $I_{n\times n} \circ (\rho_L^* J_{1\times n}) u_K$ leads to the independence of each u_i signal from all $u_{K,j}$, where $i \neq j$ and $j \in [i; n]$. Accordingly, the choosing of pairs $\rho_{L,i}^*$, $\Delta_{L,i}^*$ has no dependency on $\rho_{L,j}^*$, $\Delta_{L,j}^*$ $\forall i \neq j$; $j \in [i; n]$. Therefore, $\Delta_{L,i}$, $\rho_{L,i}$ independently from other pairs are selected.

In the case of (10.3), the equality $u_i = u_{L,i}$ is considered, such as

$$u_{L,i} = \rho_{L,i}^* u_{K,i} + \Delta_{L,i}^*, \qquad \forall i = 1 \dots n, \tag{10.7}$$

in which $\Delta_{L,i}^* \in \Lambda_{L,i}$, $\rho_{L,i}^* \in \Omega_{L,i}$. This formulation of $u_{L,i}$ describes that if $u_{K,i}$ is near to $u_{L,i}$, then connection (10.7) can be efficiently ensured by $\rho_{L,i}^*$, although if $|u_{K,i}| < \varepsilon$, where $\varepsilon > 0$ has a small value, then $\rho_{L,i}^* u_{K,i}$ has a small effect on $u_{L,i}$, as (10.7). Thus, $\Delta_{L,i}^*$ is chosen near to $u_{L,i}$. Because of these properties regarding the formulation of (10.7), a method for the selection of $\rho_{L,i}^*$, $\Delta_{L,i}^*$ is suggested. It includes two separated scenarios, which are determined by the relationship of $u_{K,i}$ and $u_{L,i}$.

In case of the first scenario, connection of $u_{L,i}$ and $u_{K,i}$ can be described as

$$|u_{L,i}| \geq \rho_{L,i,min} |u_{K,i}|, \quad \text{and} \tag{10.8a}$$

$$sgn(u_{L,i}) = sgn(u_{K,i}), \quad \text{and} \tag{10.8b}$$

$$u_{K,i} \neq 0, \tag{10.8c}$$

in which $\rho_{L,i,min} > 0$ denotes the lower boundary for $\Omega_{L,i}$. This suggests that in this scenario the signs of $u_{L,i}$ and $u_{K,i}$ are the same, and, similarly, the value of $|u_{L,i}|$ is high. Therefore, if all relations in (10.8) are ensured, the values of $\Delta_{L,i}^*$, $\rho_{L,i}^*$ are selected as

$$\rho_{L,i}^* = \frac{u_{L,i}}{u_{K,i}}, \tag{10.9a}$$

$$\Delta_{L,i}^* = 0. \tag{10.9b}$$

Some illustrations of this scenario can be found in Fig. 10.2; see non-shaded sectors.

The second scenario covers all of the other relationships between $u_{K,i}$ and $u_{L,i}$:

$$|u_{L,i}| < \rho_{L,i,min} |u_{K,i}|, \quad \text{or} \tag{10.10a}$$

$$sgn(u_{L,i}) \neq sgn(u_{K,i}), \quad \text{or} \tag{10.10b}$$

$$u_{K,i} = 0. \tag{10.10c}$$

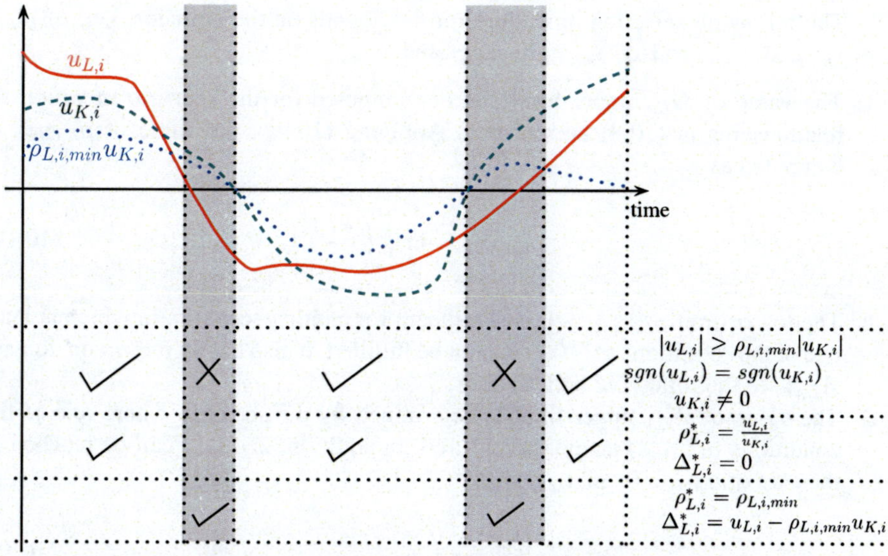

Fig. 10.2 Illustration of selection strategy for $\Delta_{L,i}$, $\rho_{L,i}$ signals

The value of $\Delta_{L,i}$ plays an important role in this scenario, such as being capable of handling zero transitions of the signals. In this scenario $\rho^*_{L,i} = \rho_{L,i,min}$ is selected, and $\Delta^*_{L,i}$ has the role to compensate the difference between the control inputs, such as

$$\rho^*_{L,i} = \rho_{L,i,min}, \tag{10.11a}$$

$$\Delta^*_{L,i} = u_{L,i} - \rho_{L,i,min} u_{K,i}. \tag{10.11b}$$

Figure 10.2 also shows illustrations of this scenario, when one of the conditions in (10.8) is fulfilled; see sectors with shading. The calculated $\rho^*_{L,i}$, $\Delta^*_{L,i}$ values are applied to the computation of $\rho_{L,i}$, $\Delta_{L,i}$; see (10.6c)–(10.6d).

10.2.2 Selection of Domains for Measured Disturbances and Scheduling Variables

Control design for variable-geometry vehicle suspension requires not only the selection of Δ_L, ρ_L but also their Λ_L, Ω_L domains. Bounds of these domains are determined by $\Lambda_{L,i} = [\Delta_{L,i,min}; \Delta_{L,i,max}]$, $\Omega_{L,i} = [\rho_{L,i,min}; \rho_{L,i,max}]$. The boundaries for domains $\Lambda_{L,i}$, $\Omega_{L,i}$ can be independently chosen from signals $i \neq j$, $j \in n$, similar to $\Delta_{L,i}$, $\rho_{L,i}$.

The following selection procedure for the bounds on the domains, i.e., $\rho_{L,i,min}$, $\rho_{L,i,max}$, $\Delta_{L,i,min}$ and $\Delta_{L,i,max}$ are suggested:

1. The value of $\Omega_{L,i}$ upper limit can be computed for the scenario in which all relationships in (10.8) are ensured. Applying (10.9a), the value of $\rho_{L,i,max}$ is computed as

$$\rho_{L,i,max} = \max\left(\frac{u_{L,i}}{u_{K,i}}\right). \tag{10.12}$$

2. The lower limit $\rho_{L,i,min} > 0$ is significant for another scenario, in which at least one of the conditions in (10.8) cannot be fulfilled. It also has an impact on domain $\Lambda_{L,i}$; see the following bullet points.
3. The value of $\Lambda_{L,i}$ upper limit is determined by the scenario where one of the conditions in (10.8) cannot be ensured. In addition, $\Delta_{L,i,max}$ can be reached in the case of $u_{K,i} < 0$ and $u_{L,i} > 0$:

$$\Delta_{L,i,max} = \max\left(u_{L,i} - \rho_{L,i,min} u_{K,i}\right). \tag{10.13}$$

   Relationship (10.13) shows that $\rho_{L,i,min}$ selection affects $\Delta_{L,i,max}$ value.
4. Likewise, $\Delta_{L,i,min}$ computation is also based on (10.11b), in which case $u_{K,i} > 0$, $u_{L,i} < 0$ inequalities are valid:

$$\Delta_{L,i,min} = \min\left(u_{L,i} - \rho_{L,i,min} u_{K,i}\right), \tag{10.14}$$

   which demonstrates that $\rho_{L,i,min}$ plays a role also in the lower bound $\Lambda_{L,i}$.

The connections between (10.12)–(10.14) show that the selection of $\rho_{L,i,min}$ affects both domains, i.e., $\Omega_{L,i}$ and $\Lambda_{L,i}$. In addition, scenario-based calculation of boundaries results in a trade-off between the impact of non-conventional control operation and the domain boundaries; e.g., if small values for $\Lambda_{L,i}$ are $\Omega_{L,i}$ selected, in that case the boundaries $\Delta_{L,i}$, $\rho_{L,i}$ can often reach domain boundaries. Therefore, u_i can frequently have a difference from $u_{L,i}$, i.e., the benefits of $\mathscr{F}(y_L)$ are not exploited. But if $\Lambda_{L,i}$, $\Omega_{L,i}$ domains have increased ranges, control input signal u_i during most of the operation time can be equal to $u_{L,i}$. Although if the learning-based control input results in degradation in the performance level, it has a higher disadvantageous impact on the achieved level of performance related to the entire controlled system.

10.3 Iteration-Based Control Design for Suspension Systems

The physical model-based representation of the variable-geometry suspension system is formed as

$$\dot{x} = A(\rho)x + B_1(\rho)w + B_2(\rho)u, \tag{10.15}$$

in which representation x is a state vector, disturbances are included in w vector and $u = \begin{bmatrix} u_1 \ u_2 \ \dots \ u_n \end{bmatrix}^T$ vector contains control input signals. Matrices of the system are denoted by $A(\rho)$, $B_1(\rho)$, $B_2(\rho)$, where $\rho \in \Omega$ contains the scheduling variables from the dynamics of the suspension, e.g., characteristics of the damper. Next, the representation of the system (10.15) is applied to design a robust LPV controller $\mathcal{K}(\rho_K, y_K)$, as (10.2) shows. The aim of LPV control synthesis is to ensure a minimum level of closed-loop performance, in which the rule of (10.5) is involved, such as $u = I_{n\times n} \circ (\rho_L J_{1\times n})u_K + \Delta_L$. Thus, the variable-geometry suspension model representation (10.15) is redefined as follows:

$$\dot{x} = A(\rho_K)x + B_1(\rho_K)w_K + B_2(\rho_K)u_K, \tag{10.16}$$

in which transformed state-space representation $\rho_K \in \Omega_K$ is constructed as $\rho_K = \begin{bmatrix} \rho \ \rho_L \end{bmatrix}^T$, $\Omega_K = \begin{bmatrix} \Omega \ \Omega_L \end{bmatrix}^T$. Similarly, $w_K = \begin{bmatrix} w \ \Delta_L \end{bmatrix}^T$ in the transformed suspension model (10.16) is constructed. The matrices are also built as

$$A(\rho_K) = A(\rho), \tag{10.17a}$$
$$B_1(\rho_K) = \begin{bmatrix} B_1(\rho) \ B_2(\rho) \end{bmatrix}, \tag{10.17b}$$
$$B_2(\rho_K) = B_2(\rho)\big(I_{n\times n} \circ (\rho_L J_{1\times n})\big). \tag{10.17c}$$

Through the LPV-based controller $\mathcal{K}(\rho_K, y_K)$, a minimum level on the performances in z_K can be achieved (Wu et al. 1996). The performance vector can be formed using u and w:

$$z_K = C_2(\rho)x + D_{21}(\rho)w + D_{22}(\rho)u. \tag{10.18}$$

Likewise in formulation (10.15)–(10.16), the formulation of performance relation (10.18) using $u = I_{n\times n} \circ (\rho_L J_{1\times n})u_K + \Delta_L$ can be transformed to

$$z_K = C_2(\rho_K)x + D_{21}(\rho_K)w_K + D_{22}(\rho_K)u_K, \tag{10.19}$$

in which the form of the matrices

$$C_2(\rho_K) = C_2(\rho)x, \tag{10.20a}$$
$$D_{21}(\rho_K) = \begin{bmatrix} D_{21}(\rho) & D_{22}(\rho) \end{bmatrix}, \tag{10.20b}$$
$$D_{22}(\rho_K) = D_{22}(\rho)\big(I_{n\times n} \circ (\rho_L J_{1\times n})\big). \tag{10.20c}$$

At last, the y_K input vector with the measurement for $\mathscr{K}(\rho_K, y_K)$ is formed depending on u, w and x, which leads to

$$y_K = C_1(\rho)x + D_{11}(\rho)w + D_{12}(\rho)u. \tag{10.21}$$

Using the same relationship on the control input (10.5), the equation on the measurement can be transformed to

$$y_K = C_1(\rho_K)x + D_{11}(\rho_K)w_K + D_{12}(\rho_K)u_K, \tag{10.22}$$

with matrices

$$C_1(\rho_K) = C_1(\rho)x + 0_{m_K\times p}\rho_L, \tag{10.23a}$$
$$D_{11}(\rho_K) = \begin{bmatrix} D_{11}(\rho) & D_{12}(\rho) \end{bmatrix} + 0_{m_K\times 2n}\rho_L, \tag{10.23b}$$
$$D_{12}(\rho_K) = D_{12}(\rho)\big(I_{n\times n} \circ (\rho_L J_{1\times n})\big), \tag{10.23c}$$

where $0_{m_K\times p}, 0_{m_K\times 2n}$ are zero matrices.

The LPV control design procedure for finding $\mathscr{K}(\rho_K, y_K)$ leads to a controlled variable-geometry suspension system, which has quadratic stability and its induced $\mathscr{L}_2$ norm from w_K to z_K has a smaller value as $\gamma > 0$. Control synthesis for solving quadratic LPV γ-performance task leads to a feasibility task of Linear Matrix Inequalities (LMIs); see Wu et al. (1996), Sename et al. (2013) and Yang et al. (2020). For control design purposes, a fine grid (e.g., lookup tables) is applied, with which an infinite number of LMI-based constraints are approximated. Thus, the parameter variable is designed through the minimization task:

$$\inf_{\mathscr{K}(\rho_K, y_K)} \sup_{\rho_K \in \Omega_K} \sup_{\substack{\|w_K\|_2 \neq 0, \\ w_K \in \mathscr{L}_2}} \frac{\|z_K\|_2}{\|w_K\|_2}. \tag{10.24}$$

Control design task (10.24) leads to the parameter-varying $\mathscr{K}(\rho_K, y_K)$ in a state-space form (Gáspár et al. 2017).

The connection between $\mathscr{K}(\rho_K, y_K)$ design and Ω_L, Λ_L selection is detailed in Sect. 10.2. Equation (10.24) shows that Ω_L, Λ_L domains in the design task are incorporated. Thus, the domains of Λ_L, Ω_L must be selected for the design of $\mathscr{K}(\rho_K, y_K)$. Nevertheless, the performance level of the LPV-based control has an impact on their selection; see (10.12)–(10.14). This interconnection motivates the iteration-based design for minimizing the difference of u_L and u, i.e., fitting maximum performance level of the non-conventional and the entire control system. Therefore, the iteration incorporates domain and intervention terms, leading to the next optimization task

$$\min_{\substack{\rho_{L,i,min}>0,\\ \rho_{L,i,max}>0}} \sum_{i=1}^{n} R_i\Big(\rho_{L,i,max}-\rho_{L,i,min}\Big)+D_i\Big(|\Delta_{L,i,max}|-|\Delta_{L,i,min}|\Big)+T_i\bar{E}_i, \tag{10.25}$$

subject to $\rho_{L,i,max} > \rho_{L,i,min}$ inequality constraint. Term $\bar{E}_i$ represents the average relative error between u_i and $u_{L,i}$.

In the optimization problem (10.25), $T_i > 0$, $D_i > 0$ and $R_i > 0$, $i = 1 \dots n$ parameters for providing balance of the terms in the cost function are also involved. Through T, the value of $\bar{E}_i$ can be scaled, i.e., the maximum level of the performance level may be influenced. For example, with high T_i values the difference between $u_{L,i}$ and u_i is small, and, consequently, the operation of the entire control system is close to the operation of the learning-based control, which is beneficial in general operation. As a result, the $\Omega_{L,i}$, $\Lambda_{L,i}$ domains are increased. However, in case of the performance reduction in the non-conventional control element, the performance level maximum related to the entire control strategy is also degraded temporarily. It leads to keeping a limitation on T. The roles of the parameters R_i, D_i are the scaling of $\Omega_{L,i}$, $\Lambda_{L,i}$ and also the facilitating of LPV controller synthesis. For example, increasing the domain of $\Omega_{L,i}$ leads to the increasing of the grid in the LPV design. Nevertheless, the large grid can lead to numerical problems in the synthesis process due to the high differences between the vertexes. Likewise, increasing $\Lambda_{L,i}$ also induces increasing robustness, i.e., conservativeness, of the control. Since high R and D values can lead to unfeasible LMI existence problems during the control synthesis, the limitation of these parameters are also required.

The suggested solution process of (10.25) optimization includes LPV control synthesis and $\Omega_{L,i}$, $\Lambda_{L,i}$ domain selection in iterative form (Németh and Gáspár 2021b).

1. In the first step, the boundaries of $\Lambda_{L,i} = [\Delta_{L,i,min}; \Delta_{L,i,max}]$, $\Omega_{L,i} = [\rho_{L,i,min}; \rho_{L,i,max}]$ for achieving large domains are selected. Initially, $\rho_{L,i,min} = \varepsilon$ with $0 < \varepsilon$ a small value and $\rho_{L,i,max}$ with a high-value are selected. The same strategy for the selections of $|\Delta_{L,i,min}|$, $|\Delta_{L,i,max}|$ is applied. It leads to a LPV control with high robustness, and the aim of the further steps is to scale these characteristics of the control system through the iteration.
2. Using (10.24), the control synthesis is performed, i.e., the LPV controller using the predefined domains is yielded.
3. Variable-geometry suspension with control operation is analyzed using different scenarios by incorporating the yielded LPV control and Λ_L, Ω_L. The performed scenarios leads to control input signal u, and its components u_K, u_L, from which $\bar{E}_i$ can be computed. Consequently, the cost in (10.25) can also be computed.

4. The results of the analysis and domain boundaries can be changed for decreasing cost in (10.25), which leads to new $\rho_{L,i,max}$, $\rho_{L,i,min}$, $\forall i = 1 \ldots n$ values. This modification induces changing of $\Delta_{L,i,min}$, $\Delta_{L,i,max}$ bounds, which can be computed through (10.13)–(10.14).
5. In this iteration process, the steps of LPV synthesis, scenario analysis and domain selection (steps 2–4) are carried out until the resulting cost in (10.25) under a predefined small value is reduced.

If the predefined minimum value of the cost cannot be reached, or the achieved performance is not satisfactory, the modification of R_i, D_i and T_i is required and the iteration process from the beginning must be carried out. For the evaluation, it is recommended to use trust-region-reflective or simplex search methods (Coleman and Li 1996; Lagarias et al. 1998), with which the cost can be effectively reduced.

References

Coleman T, Li Y (1996) An interior, trust region approach for nonlinear minimization subject to bounds. SIAM J Optim 6:418–445

Gáspár P, Szabó Z, Bokor J, Németh B (2017) Robust control design for active driver assistance systems: a linear-parameter-varying approach. Springer

Lagarias J, Reeds JA, Wright MH, Wright PE (1998) Convergence properties of the nelder-mead simplex method in low dimensions. SIAM J Optim 9(1):112–147

Németh B, Gáspár P (2021a) Ensuring performance requirements for semiactive suspension with nonconventional control systems via robust linear parameter varying framework. Int J Robust Nonlinear Control 31(17):8165–8182

Németh B, Gáspár P (2021b) Guaranteed performances for learning-based control systems using robust control theory. Springer International Publishing, Cham, pp 109–142. Deep learning for unmanned systems

Sename O, Gáspár P, Bokor J (2013) Robust control and linear parameter varying approaches. Springer, Berlin

Wu F, Yang X, Packard A, Becker G (1996) Induced L_2 norm controller for LPV systems with bounded parameter variation rates. Int J Robust Nonlinear Control 6:983–988

Yang D, Zong G, Karimi HR (2020) h_∞ refined antidisturbance control of switched LPV systems with application to aero-engine. IEEE Trans Ind Electr 67(4):3180–3190

Chapter 11
Control Design for Variable-Geometry Suspension with Learning Methods

In this chapter of this book, the method of providing performance guarantees on the variable-geometry suspension with an unconventional control agent is proposed. Two types of applications are provided, i.e., the unconventional control agent is a neural network or a direct intervention of the driver. In both applications, guarantees on the lateral dynamics are formed as the primary performance, and thus, this chapter focuses on the control design of the high level. The dynamics and control of the low level, i.e., variable-geometry suspension dynamics, are approximated through the characteristics of a simplified model.

11.1 Control Design with Guarantees for Variable-Geometry Suspension

In the design of variable-geometry suspension control, the proposed method of Chap. 10 is applied. Thus, in this section the design of the robust control and the supervisor for the specific lateral trajectory tracking problem is presented.

11.1.1 Design of the Robust Control

The control design is based on the lateral model of the vehicle:

$$J\ddot{\psi} = C_1 l_1 \Big(\delta - \frac{v_y + \dot{\psi} l_1}{v} \Big) - C_2 l_2 \Big(- \frac{v_y - \dot{\psi} l_2}{v} \Big), \tag{11.1a}$$

$$m a_y = C_1 \Big(\delta - \frac{v_y + \dot{\psi} l_1}{v} \Big) + C_2 \Big(- \frac{v_y - \dot{\psi} l_2}{v} \Big), \tag{11.1b}$$

$$v_y = \dot{y}, \tag{11.1c}$$

B. Németh and P. Gáspár, *Control of Variable-Geometry Vehicle Suspensions*, Advances in Industrial Control, https://doi.org/10.1007/978-3-031-30537-5_11

where $\dot{\psi}$ is yaw rate, v_y is lateral velocity, δ is front wheel steering angle, v is longitudinal velocity, a_y is lateral acceleration of the vehicle and y is its lateral position. Moreover, C_1, C_2 are cornering stiffness on the front and the rear axle, and l_1, l_2 are distances of the front and rear axles from the vehicle center of gravity. The dynamics is reformulated to state-space representation, such as

$$\dot{x} = A(\rho)x + B_2 u, \tag{11.2}$$

where the state vector is $x = \begin{bmatrix} v_y & \dot{\psi} & y \end{bmatrix}^T$, and the control input is $u = \begin{bmatrix} \delta \end{bmatrix}$; $\rho = v$ as scheduling variable is selected.

The control input u of the variable-geometry suspension from the candidate control input from the unconventional agent u_L and from the output of the robust controller u_K is composed as follows:

$$u = \rho_L u_K + \Delta_L; \tag{11.3}$$

see (10.5). Thus, in this application example $n = 1$ due to the single control input of the system. The transformation of (11.2) through (11.3) results in the system representation

$$\dot{x} = A(\rho) + B_2 \Delta_L + B_2 \rho_L u_K. \tag{11.4}$$

The primary, i.e., the safety performance of the system is to guarantee the limitation of the lateral error of the vehicle from the centerline of the road:

$$z_1 = y_{ref} - y; \qquad |z_1| \rightarrow min, \tag{11.5}$$

where y_{ref} is the reference lateral position for the vehicle. Moreover, the limitation of the steering angle is requested to avoid the unwanted effect of actuator saturation, which leads to further performance:

$$z_2 = \delta; \qquad |z_2| \rightarrow min. \tag{11.6}$$

The performance vector $z_K = \begin{bmatrix} z_1 & z_2 \end{bmatrix}^T$ through the state space Eq. (11.1) can be expressed as

$$z_K = C_2 x + D_{21} y_{ref} D_{22} u, \tag{11.7}$$

which can be reformulated through (11.3); see (10.19)

$$z_K = C_2 x + D_{22} w + D_{22} \rho_L u_K, \tag{11.8}$$

where $w = \begin{bmatrix} y_{ref} & \Delta \end{bmatrix}^T$. Similarly, the formulation of measurement $y_K = y_{ref} - y$ is expressed as

$$y_K = C_1x + D_{11}w + D_{12}\rho_L u_K. \tag{11.9}$$

The control-oriented state-space representation of the system from the dynamics, performances and measurements of the system is composed as

$$\dot{x} = A(\rho) + B_2\Delta_L + B_2\rho_L u_K, \tag{11.10a}$$

$$y_K = C_1x + D_{11}w + D_{12}\rho_L u_K, \tag{11.10b}$$

$$z_K = C_2x + D_{22}w + D_{22}\rho_L u_K. \tag{11.10c}$$

The system (11.1) is parameter-dependent with disturbance vector w, whose impact on the performance vector z must be minimized. Therefore, the robust LPV design method of the control synthesis is selected, which is able to provide the stability of the closed-loop system together with disturbance attenuation Bokor and Balas (2005). Scaling of disturbances and performances is requested for the robust LPV design, and, thus, the plant (11.10) is augmented with weighting functions; see Fig. 11.1. The system (11.10) is represented as $G(\rho, \rho_L)$ and the controller is $K(\rho, \rho_L)$. The reference signal y_{ref} is scaled with the function $W_{ref} = \frac{y_{ref,max}}{T_{ref}s+1}$. It represents that for a steady-state scenario the maximum of the reference signal is $y_{ref,max}$, and, moreover, its variation can have dynamics with T_{ref} time constant. Performance z_1 is also scaled with a transfer function to represent the allowed dynamics of the tracking error. In the function of $W_{z_1} = \frac{1/e_{max}}{T_e s+1}$, e_{max} represents the maximum lateral error in steady state and T_e represents the time constant of the tracking error dynamics. Moreover, in the augmented plant the further weights are selected to be scalar values. $W_\Delta = \Delta_{max}$ is related to the bound of Δ, and W_n is the weighting function of the sensor noise. Finally, $W_{z_2} = \frac{1}{\delta_{max}}$ scales the control input, whose maximum is allowed to be δ_{max}.

The optimization problem of the control design is based on the method of Sect. 10.3. Thus, it is necessary to choose the parameter-varying controller $K(\rho, \rho_L)$ in such a way that the resulting closed-loop system is quadratically stable and the induced $\mathscr{L}_2$ norm from the disturbance and the performances is less than the value γ. The minimization task is the following:

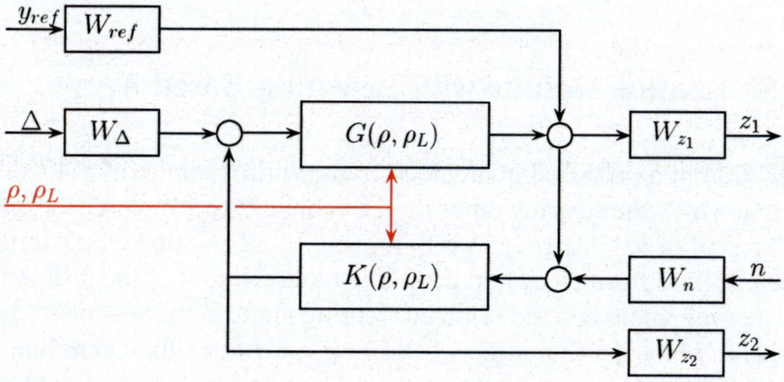

Fig. 11.1 Illustration of the augmented plant

$$\inf_{K(\rho,\rho_L)} \sup_{\substack{\rho \in \boldsymbol{\rho}, \\ \rho_L \in \boldsymbol{\rho}_L}} \sup_{\substack{\|w_K\|_2 \neq 0, \\ w_K \in \mathcal{L}_2}} \frac{\|z_K\|_2}{\|w_K\|_2}, \tag{11.11}$$

where $\boldsymbol{\rho}$ covers the bound of ρ. The result of the optimization process is $K(\rho, \rho_L)$, whose output during the operation of the control is u_K.

11.1.2 Forming Supervisory Algorithm for Variable-Geometry Suspension

For the formulation of the supervisor, the algorithm in Sect. 10.2 is applied. The core of the method is that u must be in the limited surrounding of u_K and $|y_{ref} - y|$ must be smaller than e_{max}. If these conditions are guaranteed, $u = u_L$ control input is used and ρ_L, Δ_L are selected by the rule of (10.9a). In case these conditions are not guaranteed, the selection of u, ρ_L, Δ_L is as follows:

- If $u_L < \rho_{L,min} u_K + \Delta_{L,min}$, i.e., u_L is out of the limited lower bound of u_K, then u is limited through the selection of $\rho_L = \rho_{L,min}$, $\Delta_L = \Delta_{L,min}$. It guarantees the robustness of the closed-loop systems, because the difference between u and u_K is inside of the predefined range in the LPV control design.
- If $u_L > \rho_{L,max} u_K + \Delta_{L,max}$, i.e., u_L is out of the limited upper bound of u_K, then u is limited by the upper bound for guaranteeing robustness. In this case $\rho_L = \rho_{L,max}$, $\Delta_L = \Delta_{L,max}$ are selected, which results in $u = \rho_{L,max} u_K + \Delta_{L,max}$.
- If the condition on the relation of u_K, u_L is not violated, i.e., the condition on $|y_{ref} - y| < e_{max}$ is not guaranteed, the selection of $\rho_L = 1$, $\Delta_L = 0$ is used. It results in the tracking performance level of the LPV controller being achieved, which is able to guarantee the reduction of the lateral tracking error.

Through the cooperation of the robust LPV controller and the supervisory algorithm with the previous rules, the primary performance, such as the limitation of the tracking error, can be guaranteed.

11.2 Simulation Results with Learning-Based Agent

In this section, the proposed guaranteed framework is applied for a vehicle control problem, in which the steering input for the vehicle is recommended by a neural network. The goal of vehicle control is to imitate driver steering characteristics, which have been learned from measured data. The minimization of the difference between the real steering angle and the imitated steering angle is the secondary performance requirement, while the limitation of tracking error from the centerline of the road is the primary performance specification. In some papers, the problem of imitation learning for autonomous vehicles has already been studied. For example, an end-to-end learning method in Codevilla et al. (2018) has been proposed, where a vision-based steering control through conditional imitation learning has been achieved. Paper Kebria et al. (2020) has proposed a general framework for the selection of convolutional neural network parameters in the case of deep imitation learning problems. Moreover, in Pan et al. (2020) imitation learning has been applied for agile autonomous driving, which provides special challenges under extreme driving situations. In the context of variable-geometry suspension, the method of imitation learning has not been used.

The advantage of imitation learning is to use neural networks, by which the characteristics of the driver can be achieved. Through the fitting of neural networks, it is possible to handle control problems, in which nonlinearities have high impacts. In the structure of the neural networks, several layers are found, i.e., input layer, hidden layer and output layer. A layer consists of neurons, which are built up by activation functions and weights. The selection of layer number and neuron number is not a trivial task, because of their impact on the effectiveness of the training process. A general technique to select these parameters is the k-fold cross-validation technique Demut et al. (1997), and for the selection of the number of neurons a method can be found in Xu and Chen (2008).

In this simulation example, a medium-sized passenger vehicle in CarMaker is selected, in which the intervention dynamics of variable-geometry suspension through a time delay and a transfer function is approximated. The driving of the vehicle has been performed through the driver model of CarMaker, and, thus, data on the vehicle motion and the steering intervention has been collected. For the fitting of the neural network, the following input signals on the neural network have been used:

- actual lateral error of the vehicle,
- actual longitudinal velocity of the vehicle and
- actual and forthcoming curvature of the road in 1 s horizon with 0.1 s consecutive steps.

Moreover, the output of the neural network is the generated steering angle of the driver, which must be fitted to the steering angle of the measured dataset. In this example, 2 hidden layers with 20 neurons in each layer have been selected. The learning process has been performed through the Levenberg–Marquardt algorithm; see Demut et al. (1997). Figure 11.2 illustrates an example of the results of the training process. It can be seen that the error of the driver steering wheel intervention and the output of the neural network is very small, and it results in negligible steering difference on the front wheels.

The following figures show the result of the simulation using the performance guaranteed control framework. The path tracking capability of the vehicle is shown in Fig. 11.3. The course of the vehicle is illustrated in Fig. 11.3a. It can be seen that through the proposed control system, the automated vehicle is able to perform the track. The lateral error during the route of the vehicle can be seen in Fig. 11.3b. The results show that the control system is able to guarantee the limitation of the lateral error, since e_{max} has been set to 4 m during the robust LPV control design.

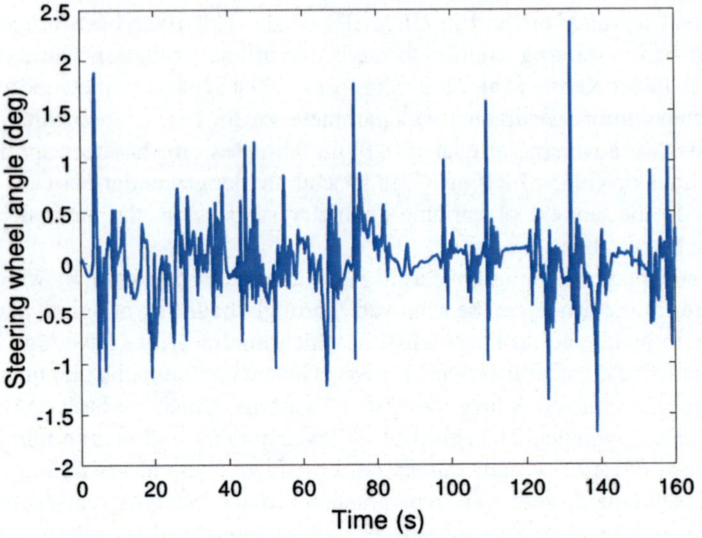

Fig. 11.2 Result of steering wheel fitting error

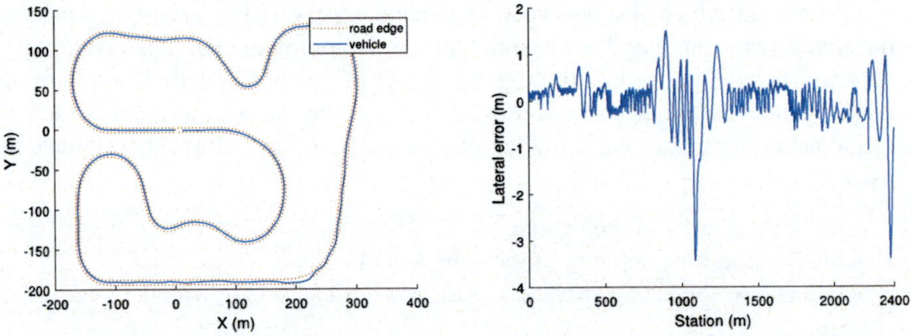

(a) Course of the vehicle (b) Lateral error from the centerline

Fig. 11.3 Motion of the vehicle on the track

Details on the control intervention are presented in Fig. 11.4. Signals δ and δ_L are found in Fig. 11.4a. The results show the effectiveness of learning driver intervention, because in most of the simulation $\delta = \delta_L$ is achieved. The value of ρ_L and Δ_L are illustrated in Fig. 11.4b–c.

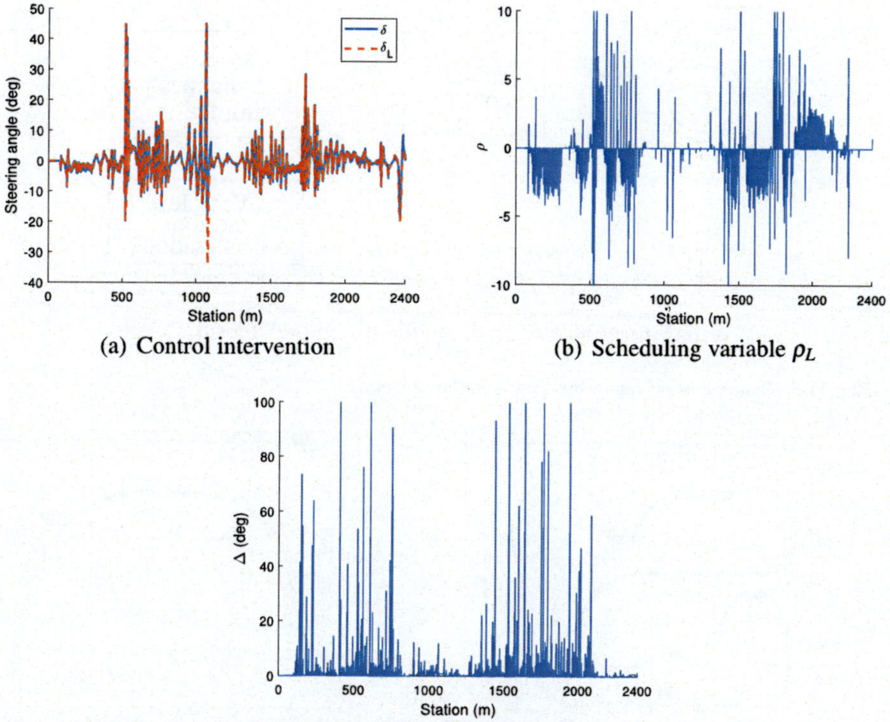

(a) Control intervention

(b) Scheduling variable ρ_L

(c) Uncertainty Δ_L

Fig. 11.4 Details on the control intervention with neural network

11.3 Simulation Results with Driver-in-the-Loop

In this section, another application of trajectory tracking control in a guaranteed framework is presented. Despite the previous example, in this case the unconventional agent is not an automated control element, but the steering intervention of the driver. The example illustrates that the use of the design framework is independent of the structure of the unconventional agent.

The architecture of the driver-in-the-loop simulation setup is illustrated in Fig. 11.5. In the scheme, the driver rotates the steering wheel, which has force feedback functionality, and, simultaneously, the motion of the vehicle is visualized in real time by CarMaker. The generated steering wheel angle is the candidate control input δ_L for the vehicle control, which covers the supervisor and the robust LPV controller. It also measures the tracking error from CarMaker. The vehicle control in the simulation setup is implemented in MATLAB® and Simulink®. The computed δ steering angle is the input for CarMaker, which facilitates the motion of the vehicle.

In the rest of this section, three simulation scenarios are shown to illustrate the effectiveness of the control. In the first simulation, the driver is in the loop and the

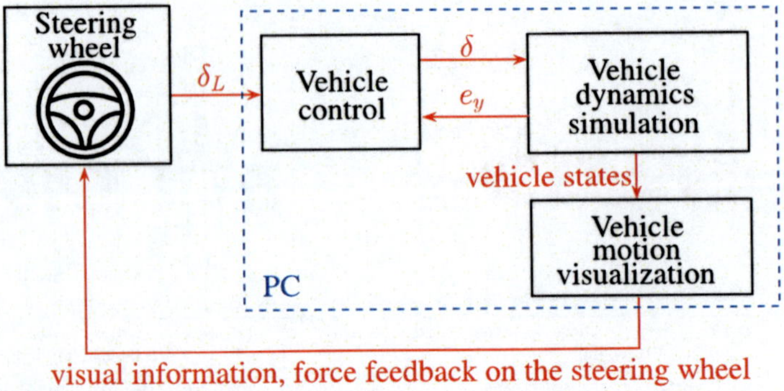

Fig. 11.5 Scheme of driver-in-the-loop simulation setup

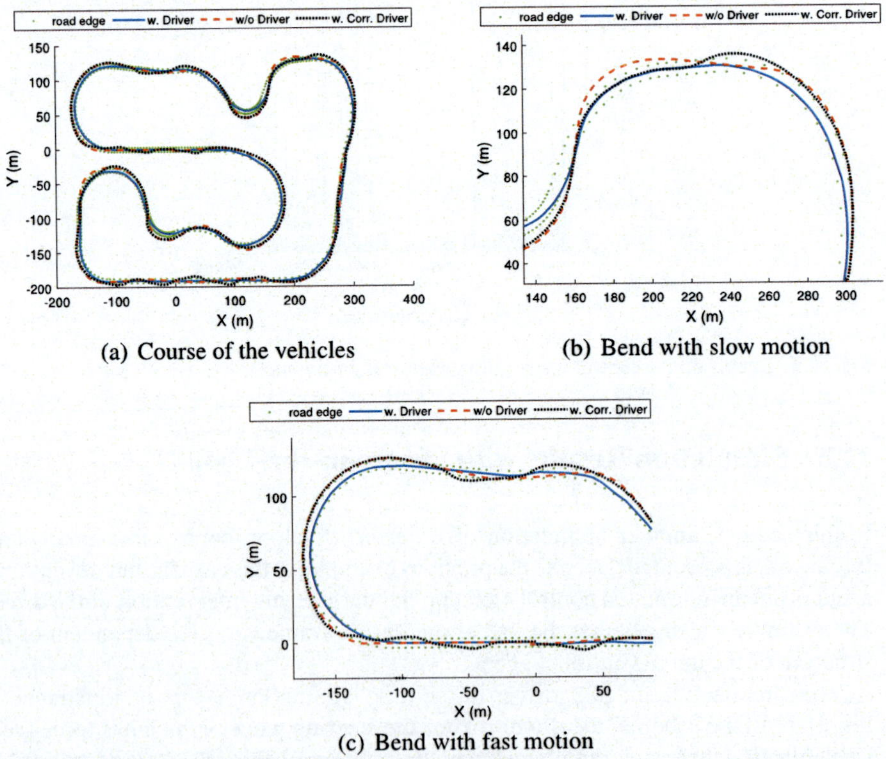

Fig. 11.6 Path of the vehicles in the three scenarios

goal is to move the vehicle on the track (*w. Driver*). Second, the driver is in the loop, but $\delta_L \equiv 0$ during the entire scenario, which imitates a sleeping/inactive driver (*w/o Driver*). In the third scenario, the goal of the driver is to corrupt tracking performance, i.e., it is selected steering wheel angle with which the position of the vehicle is as far as possible for the centerline (*Corr. Driver*).

The motion of the vehicle on the road in the different scenarios is illustrated in Fig. 11.6. Figure 11.6a illustrates the entire track, in which it can be seen that the vehicle does not move far from the path in the case of the three scenarios. Two illustrative examples are shown in Fig. 11.6b–c. In Fig. 11.6b, a sharp bend with slow motion of the vehicles is shown. It can be seen that with the driver, the bend inside of the path is performed. Although in the two further cases the vehicle slightly left the path, through the intervention of the automatic control the performance degradation is limited. Similarly, in Fig. 11.6c the bend with fast motion of the vehicle is shown. Although the corrupted driver tries to leave the path through the oscillating motion of the vehicle (see road section at $Y = 0$), the performance level on maximum tracking error is not violated.

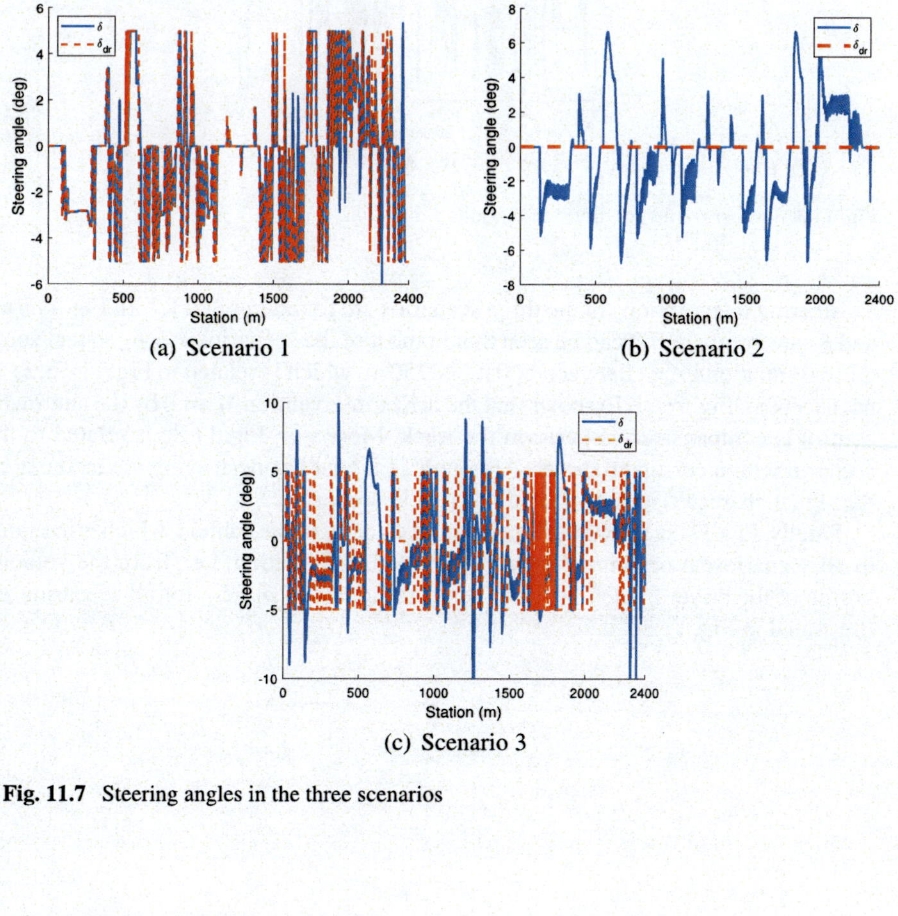

(a) Scenario 1 (b) Scenario 2

(c) Scenario 3

Fig. 11.7 Steering angles in the three scenarios

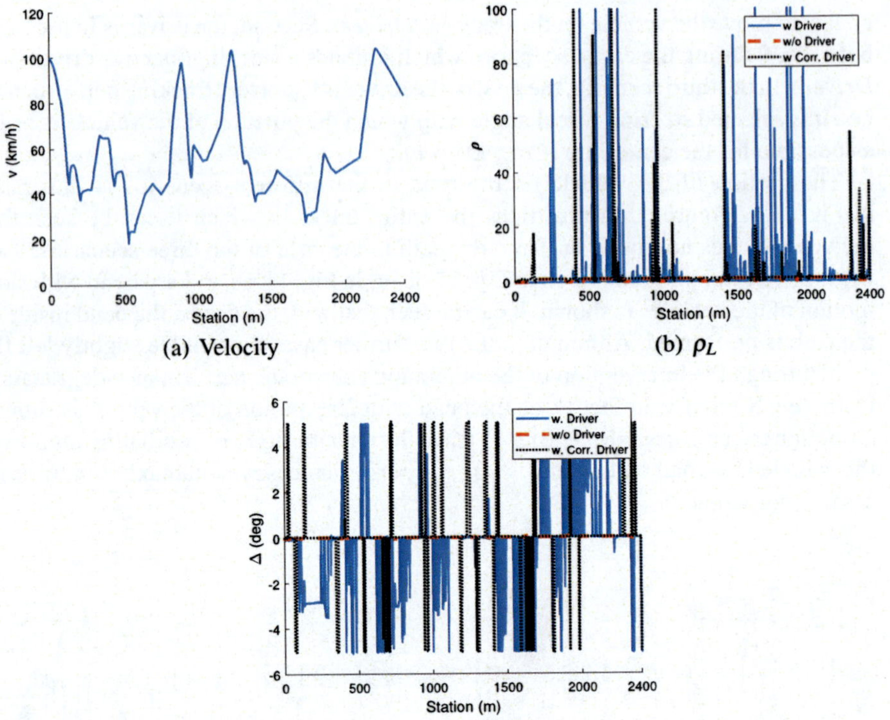

(a) Velocity

(b) ρ_L

(c) Δ_L

Fig. 11.8 Further results for driver-in-the-loop simulations

Steering interventions in the three scenarios are found in Fig. 11.7. In Fig. 11.7a δ and δ_L are compared. It can be seen that in most of the scenarios $\delta = \delta_L$, but at some critical situations, e.g., between 600 m . . . 750 m, which is related to Fig. 11.6b, δ_L is not accepted. Figure 11.7b shows that the driver intervention $\delta_L \equiv 0$ by the automatic control is compensated to perform the track. Moreover, Fig. 11.8c is related to the scenario with a corrupted driver. Although δ_L signal has destroying characteristics, the signal characteristics of δ are close to δ in scenario 2.

Finally, Fig. 11.8a shows the longitudinal velocity of the vehicle, which is the same in all scenarios. It originated from the previous application, i.e., from the velocity profile of the driver model in CarMaker. The signals of ρ_L, Δ_L for all scenarios are illustrated in Fig. 11.8b–c.

References

Bokor J, Balas G (2005) Linear parameter varying systems: a geometric theory and applications. In: 16th IFAC world congress, Prague, pp 1–6

Codevilla F, Müller M, López A, Koltun V, Dosovitskiy A (2018) End-to-end driving via conditional imitation learning. In: 2018 IEEE international conference on robotics and automation (ICRA), pp 4693–4700

Demut H, Hagan M, Beale M (1997) Neural network design. PWS Publishing Co

Kebria PM, Khosravi A, Salaken SM, Nahavandi S (2020) Deep imitation learning for autonomous vehicles based on convolutional neural networks. IEEE/CAA J Autom Sinica 7(1):82–95

Pan Y, Cheng CA, Saigol K, Lee K, Yan X, Theodorou EA, Boots B (2020) Imitation learning for agile autonomous driving. Int J Robot Res 39(2–3):286–302

Xu S, Chen L (2008) A novel approach for determining the optimal number of hidden layer neurons for fnn's and its application in data mining. In: 5th international conference on information technology and applications (ICITA 2008), pp 683–686

Index

B. Németh and P. Gáspár, *Control of Variable-Geometry Vehicle Suspensions*, Advances in Industrial Control, https://doi.org/10.1007/978-3-031-30537-5